NATIONAL PARKS OF LATIN AMERICA

LANDSCAPES, CHALLENGES, AND ROUTES TO ENVIRONMENTAL CONSERVATION

PHOTOGRAPHY
LUCIANA NAPCHAN

NATIONAL PARKS OF LATIN AMERICA

RESEARCH AND TEXT EDITOR
WALTER BEHR

Concept, production, and contents Luciana Napchan and Walter Behr
Photography and captions Luciana Napchan
Research and text editor Walter Behr
Final text Ana Augusta Rocha
Technical advisory Rozely Ferreira and Claudio Maretti
Art direction Rudi Böhm
Graphic design Márcia Signorini and Mário Ferraz Jr.
Graphic production Márcia Signorini
Translation and proofreading coordenation Marina Gonzalez
English translation Peter Musson
English proofreading Paul Ludlow, Paul Dale, Claudia Vieitas, Aline Matos Gurgel, and Regina Trindade
English technical revision Claudia Vieitas and Paul Dale
Printed by Gráficos Burti, São Paulo, Brasil.

Published simultaneously in Portuguese, Spanish, and English

Sponsors Petrobras (main sponsor), Natura, Burti

This edition has the support of Suzano Papel e Celulose, Advance Turismo, Lan Chile, Half Dome, Furtado & Schmidt, Nikon T. Tanaka, Fujifilm do Brasil, and IBAMA

Luciana Napchan
www.luciananapchan.com

Walter Behr
walterbehr@hotmail.com

Dados Internacionais de Catalogação na Publicação (CIP)
(Câmara Brasileira do Livro, SP, Brasil)

National Parks of Latin America / Photography Luciana Napchan ; research and text editing Walter Behr; [English translation Peter Musson]. -- São Paulo : Ed. dos Autores, 2005.

Obra em 3 v. : português/inglês/espanhol
Bibliografia.
ISBN 85-905070-3-3

1. Conservação da natureza 2. Ecossistemas 3. Gestão ambiental 4. Parques e reservas nacionais - América Latina 5. Parques e reservas nacionais - Fotografias 6. Proteção ambiental I. Napchan, Luciana. II. Behr, Walter.

05-0011 CDD-333.78316098

Índices para catálogo sistemático:

1. América Latina : Parques nacionais : Conservação e proteção 333.78316098
2. Parques Nacionais : Conservação e proteção : América Latina 333.78316098

LEI DE INCENTIVO À CULTURA
MINISTÉRIO DA CULTURA

After visiting a region of exceptional beauty in the surroundings of a national park, the desire was born to give back to nature the sensation of peace and harmony that it had given us.

The idea was that by sharing the beauty and importance of environmental conservation, we would, in some way, be reaching the seedbed of this process, that is, the human beings.

A national park is an important category of protected area. In addition to protecting fundamental parts of ecosystems and their landscapes, it is also a field for the development of research, education, and environmental tourism.

That was how the idea arose for producing this book, demonstrating examples of successful environmental conservation in the Latin American National Parks.

Our initial challenge was to choose, out of hundreds of protected areas, eight national parks that would represent different ecosystems in Latin America and that would demonstrate, either within their boundaries or in their surrounding areas, successful environmental conservation projects and programs that could be replicated in other places of this region.

Documenting these experiences and these natural heritages convinced us that the countries of Latin America have much to share among them.

The environmental conservation projects portrayed in this book are fascinating and diverse. By overcoming difficulties and challenges within similar socioeconomic environments, they deserve to be considered as "conservation treasures" to Latin America.

Due to my passion for nature, I feel the need to cry out through my photography, as trees and animals are unable to do it for their own conservation. They simply die and disappear. *Luciana Napchan*

Society needs to recognize that national parks are common areas, benefiting everyone. We need to feel responsible for these areas and take a greater part in their protection. Otherwise, our natural areas will continue to be destroyed. *Walter Behr*

NATIONAL PARKS OF LATIN AMERICA

Protected areas are becoming islands.

And now, after almost two centuries of incessantly pillaging nature, we reach out, like the shipwrecked, toward it.

Egocentric as we are, in this short period of time called "modernity", we have decreed that the Earth should be more human than anything else. The result is that we now have a red book, within whose white pages we see an immense blacklist of species that have disappeared or are about to disappear. Dominate – was the order of the day, according to the "progress" statement. Letting live was purely secondary, or perhaps superfluous.

Now, surprised, we have discovered how empty it feels to live without this lost biodiversity.

It empties us of a future.

While there were forests, there was freshness, clean water, and the secrets of plants to discover. Few remember, but a good part of the medicines currently known to us originated in tropical forests, and from the study of their plants. And what are we doing with these plants?

We are eliminating them before we have even studied them. Over half the planet's tropical forests have already been destroyed and continue to disappear with each passing day. Each species that falls could be taking with it the chance for advances in science and in medicine.

The climate of the planet also faces an uncertain future. The purity of its waters is lost. Year after year, the lands that provide our food become arid, less fertile, and disappear…

The extinction of species also empties us in an ethical way. Why believe that we humans should have a greater right to life than any other animal or plant, which probably existed on the planet well before us?

And what will we say to our descendants?
Say we are sorry? Say that we sort of overdid it? Point out to them in an affable voice, but with trembling fingers, the few protected areas – currently less than 12% of the planet, according to The World Conservation Union (IUCN) as a minimal, even if doubtful, proof of our sanity?
Will we say to them, as if telling a fairy story, that these are treasure islands?

There is still time to leave a better heritage for future generations. With more ingenuity in finding environmental solutions, with a greater feeling of identification with all forms of life, and, above all, with an understanding of the world that includes perceiving the nature in each one of us, we will intrinsically understand that we are a part of and dependent on nature.

Perhaps we will be able to see that the exaggerated consumption of our natural resources and of material goods will inevitably lead to a poverty of life.

Perhaps the last world's isolated center of the diversity could expect more of us.

And instead of contenting ourselves in having parks and reserves surrounded by civilization and problems on all sides, we could, by going against the tide of destruction: extend corridors for fauna and flora to interconnect protected areas, stretch boundaries, increase sustainable development experiences and tools, involve local people, exchange experiences and successful ideas, stimulate research and partnerships, and widen environmental education and involvement at all levels of society.

These are the ideas and they represent the ideal of this book.

We must build bridges and join – and cross limits, whether of countries, cultures or conservation areas, so as to share solutions, hopes, and routes – routes to conservation.

Because we know that mankind must urgently leave the sphere of thoughts, ideas, and intentions, for that of practical reality. We need to see our ethics and our hopes consolidated in action. Sustainability cannot remain merely an idea on paper or in our minds.

It has taken us five years to see this book published.

Perhaps it will take another fifty to see the ideas it promotes widely practiced.

The important thing is to make a start, by turning the pages and writing a new story.

Protection of the environment has, without a doubt, always been one of Petrobras's priorities. Throughout its 51-year existence, the company has become a partner of institutions that have set the pace both in Brazil and abroad for the serious and innovative work they have undertaken. The Tamar Project, working for the conservation of sea turtles, is one of them.

The book *National Parks of Latin America* is another fine example of this partnership. And the principal beneficiary is you. On each page and in every photograph, the reader will have the opportunity to find out about what has been done to preserve ecological sanctuaries in eight national parks in Latin America.

Over the last four years, Petrobras has invested around R$ 6 billion in the environmental security of its installations and another R$ 18 million in sponsoring environmental projects.

The Petrobras's environmental programme "Petrobras Ambiental", launched in 2003, represents a new level in this process and foresees investments of R$ 40 million, in its first two years, in projects chosen by public selection.

In October 2003, the company became a signatory of the United Nations Global Pact. Among the principles of this initiative are the promotion and dissemination among its workforce and its thousands of suppliers and shareholders of the concepts and precepts of social and environmental corporate responsibility.

Petrobras is fully aware that respecting the environment is an act of exercising one's civic responsibility, benefiting not only Brazil, but also the other nations of our great sanctuary: Planet Earth.

BR PETROBRAS

Natura believes that life is a chain of intertwined relationships. They are so connected that the respect to one of its parts is reflected in the whole, and the care devoted to one part is also felt in the whole.

The company also believes that it is urgent to protect nature, and this has necessarily to go through a review of values and attitudes.

To contribute for this change, Natura strives to strengthen the belief that it is possible to combine a progress that does not destroy nature to development and full achievement of human values. The sharing of this belief is a step required to build a better world.

This book has the special importance of showing consistent and inspiring examples of how to take care of nature. Examples that speak of the management of national parks in Latin America and that can be copied elsewhere in the world. National parks – as well as other conservation units – are wild life reserves that have become increasingly important for mankind. In addition to preserve natural environments, they contain the resources required to restore nature.

By supporting this book, Natura reiterates its 35 years of commitment to sustainable growth in Brazil, in Latin America, and in the world. The dissemination of knowledge works in this direction: to be knowledgeable in order to protect.

natura

This year, Burti is celebrating its 30th anniversary.

These have been 30 years of hard work, fascinating challenges, great successes, and impressive triumphs. During these 3 decades, we have witnessed the rapid growth of Brazil, the evolution of markets and of the mass consumer society, as well as the explosive advent of globalization.

Over this time, however, it has become increasingly clear that our planet is very fragile and that, as this relative progress and technological evolution continue, signs are appearing that we must rethink our impact on natural resources and consider how to achieve sustainable growth. We must recognize our natural riches and tenaciously conserve them for future generations. Imbued with this spirit, we have joined hands with other partners to support the *National Parks of Latin America* book project. We believe that this initiative will help make known the work of innumerable people who contribute, directly and indirectly, to the preservation of the planet Earth and that it will reveal to us, the readers, a small part of the beauty that is available to us but that goes unnoticed amid our daily routines. I invite you to share this journey and to explore new frontiers, discovering the horizons of a continent that is great by nature.

Gráficos Burti

4
7
8
1
3
6
5
2

8 eight chapters
eight Latin America regions
eight national parks
eight badly known paradises

1 One of the Earth's greatest biodiversities. Manu National Park.

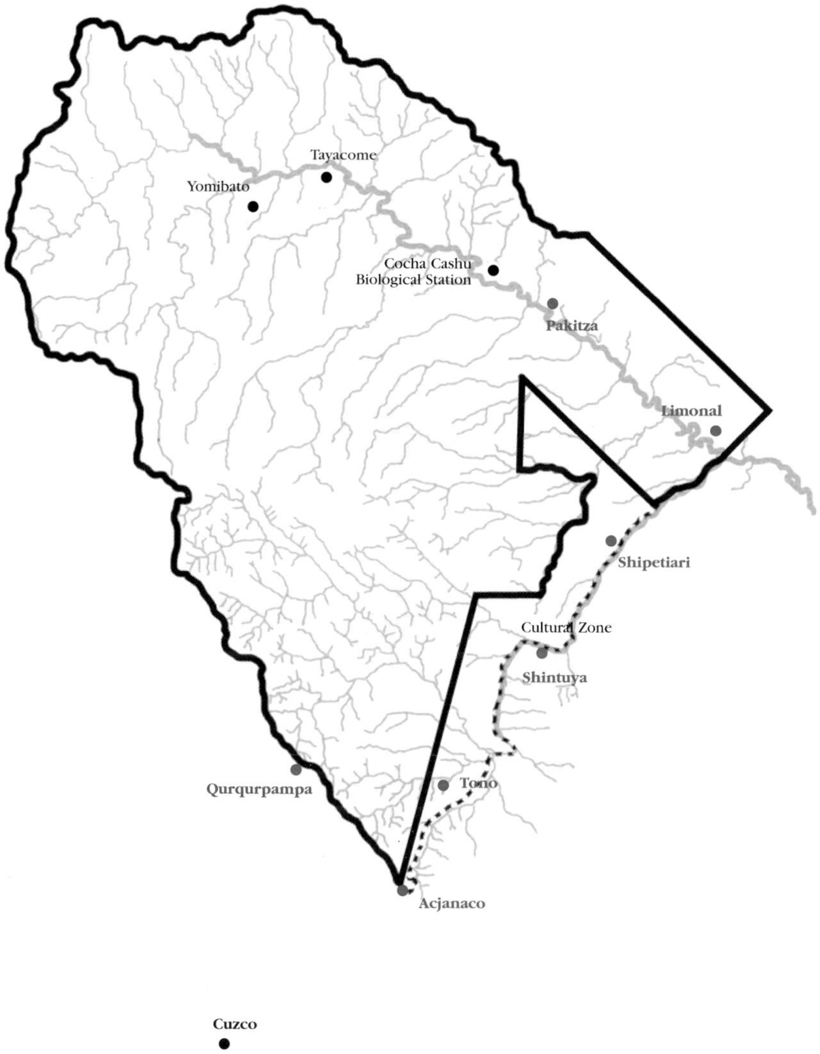

Location Eastern slopes of the Peruvian Andes, 100 km from Cuzco.
Area 1,692,137.26 hectares.
Creation Declared a national park in 1973, a Biosphere Reserve in 1977, and a Natural World Heritage Site in 1987.
Ecosystem The Manu covers three different ecological zones: the Puna, found at high altitudes where vegetation is scarce; the high-altitude cloud forest, on mountain slopes where moisture and the number of endemic species are extremely high; and the Amazon Forest, home to one of the world's richest biological diversities.
Climate The climate is temperate in the high Andes, where rainfall is extremely low; subtropical on the mountain slopes, with up to 6,000 mm of rainfall per year, and equatorial in the Amazon area of the park, where rainfall can reach 2,100 mm per year.
Relevant features Ecotourism, scientific research, and international relations.

1 Manu National Park. One of the Earth's greatest biodiversities.

The boat goes slowly up the Manu River, allowing the tourists aboard to film and admire the typically Amazon landscape of towering trees and a muddy river, with the occasional movement of birds or animals on its banks. Around a bend in the river, flowing against the course of the river and the history of the 20th century, three immense canoes are coming downstream. Aboard them are at least a dozen of the feared Yaminahua warriors, with whom contact was only initiated a few years ago. Naked, except for a minute loincloth and the bows and arrows slung over their shoulders, they seem a vision out of the past.

There is a tense moment: the tourists lower their cameras and also, slowly, their eyes. The pilot further reduces speed as if to avoid offending the silence of the canoes. The forest seems immobile. Only the Indians face the situation calmly, turning their heads to stare at these overdressed usually chattering people, who now appear to have suddenly lost their tongues, their gestures, their very way of being.

Another bend soon appears and the two opposing tribes return to their courses, moving apart. At the scene of the meeting, the waters run smoothly and muddy once more. The trees stretch out, reaching for the sun. The sounds of the forest return. It is as if nothing had happened.

The Manu National Park lies mainly within the Amazon basin, and for this very reason carries within its core the characteristic ability of tropical forests, when left alone, to repair and re-invent themselves to the smell of the rains and the moisture.

Even though of difficult access, hidden behind the escarpments of the Andes Mountains, these lands have already been the scene of numerous encounters between opposing forces: the Incas against the native people, the Spanish colonizers against the Incas, the Spanish against the Indians, and the rubber barons against everyone, including the trees. The region of the Manu has witnessed all these struggles, and, to a lesser or greater degree, has been the subject of aggression over the centuries. Curiously, after a while, it seems to have emerged unscathed.

In 1967, an English major, Ian Grimwood, was charged by the Peruvian government to find a suitable area for the creation of a national park. He was ready to return to England, judging the mission a failure, when he met the naturalist Celestino Kalinowsky – considered to be the "father" of the Manu – and accepted an invitation to explore the area. He returned filled with the dream of making it a park and the certainty of never having seen such a concentration of wildlife in any other part of the Amazon.

The Manu National Park is one of the largest reserves of tropical forest in the world, with an extremely high degree of biodiversity within its 1,692,137.26 hectares. A national park since 1975, a declared Biosphere Reserve by UNESCO in 1977, and a World Heritage Site since 1987.
Manu has a lot of titles and complexities.
The same can be said of its landscapes.

The Manu National Park is unique in the world for covering such a wide range of ecosystems. It starts in the high Andes, specifically at 4,020 m, in a totally Andean landscape known as the Puna, reminiscent of the kind of stunted vegetation found in the European tundra and where golden grasses sway in the predominantly chill winds. This half-frozen Andean Peru is only 100 km from the ancient Inca capital of Cuzco and the ruins of Machu Picchu. Nevertheless, only a few kilometers from its eastern border, going down toward the Amazon basin, the Manu's features are radically altered: as we descend the high Andes, we are plunged into a damp and mist-shrouded world: that of the high-altitude cloud forest. This is a mysterious area, composed of dense vegetation consisting of low trees with perennial leaves, extending between the altitudes of 3,500 and 1,000 m. Poorly studied, the cloud forest is known to have a high degree of endemic plant species.

A few more hundred meters down toward the Manu River and the visitor will abruptly enter a third kingdom: that of the lowland Amazon basin, that of towering trees, countless rivers, and one of the most varied biodiversities on Earth. The forest is literally teeming with life. The Amazon portion of the Manu National Park is renowned for the immense number of animals to be found.

The skies are decorated with birds, especially exotically colored macaws that feed on minerals found in the riverbanks. Over 1,000 bird species have been counted in the Manu, around 10% of all the planet's species. The concentration of monkey species (more than 12) is also high, ranging from the very small, such as the pygmy marmoset *(Cebuella pygmaea)*, weighing just 120 grams, to the woolly monkey *(Lagothrix lagotricha)*, at up to 11 kilos. Another special inhabitant is the giant river otter, *(Pteronura brasiliensis)*, extinct in many of South America's rivers, but found in the Manu in significant number.

The ease with which the fauna can be observed in the Manu was fundamental to the history of the park, attracting world-famous researchers to the area. So much so, that already by 1970, there was a research station within it founded by two German biologists, which would later be called the Cocha Cashu Biological Station, from whose research over 300 scientific papers have already been published.

The involvement of these researchers over the decades has proved to be fundamental in publicizing and strengthening the park as an institution – both in creating an immense body of knowledge that has helped in its management, as well as in making the Manu a heritage site known and admired around the world. The high profile brought by researchers has been the driving force behind the involvement of others, a guarantee of conservation, and the first step in establishing partnerships. Such was the case with the American Ornithologist Charles Munn who, after producing highly detailed scientific research on the park that was seen on various TV networks and published in internationally renowned magazines, created various ecological tourism programs in the areas surrounding the park. With the help of the Peru Verde Non-Governamental Organization (NGO), whose structure he helped set up, the ecotourism programs have brought a new prospect of subsistence for the local communities, which have become important allies in environmental conservation. More recently, the program has been broadened to include the local indigenous reserves.

While raising funds and promoting the training of the indigenous population for this new enterprise, Munn oversaw the construction of a tourist lodge to the west of the park, more specifically in the Timpia community. Within just six months of its inauguration, and working at an average occupation of only 6%, the lodge already generated – without alterations to the environment – more resources than the traditional coffee plantations. The occupation rate (and income) has been growing ever since, generating an important alternative source of income to the community.

Another point that has projected the Manu as a protected area attracting funding, tourists, and other support, was its elevation to Biosphere Reserve status by United Nations Educational, Scientific and Cultural Organization (UNESCO) through the organization's Man and Biosphere (MaB) program. MaB, launched in 1968 during a Conference on Conservation and Rational Use of the Biosphere, faces up to one of the most urgent problems challenging mankind today: how to reconcile economic development with the preservation of biodiversity and with improving quality of life for this and future generations. It is a program, managed by scientists and environmentalists from around the world, that details the concepts and prerequisites of environmental protection that must be implemented in practice. The process begins when the governments involved, together with NGOs, and civil society, lobby the MaB for designation of a particular area as a Biosphere Reserve. Whatever the outcome, by requesting Biosphere Reserve status from the MaB, all parties take on serious commitments to conservation, sustainability, and social justice in the areas involved. It can be likened to a stamp of quality added to a region that both opens doors and brings responsibilities. In the case of the Manu, in addition to the park, a contiguous area, in which a cultural zone and a reserved zone have been set aside, is also part of the Biosphere Reserve.

The cultural zone is similar to the Brazilian model of an extractive reserve, ensuring the sustainable use of natural resources by the people living within it. The reserved zone, the aptitudes for which were for a long time undecided, is now open to visitors, and has recently been incorporated into the area of the park. The three areas total 1,881,200 hectares and are an incentive for other areas adjacent to the park to be acquired for preservation, whether by NGOs or by private owners, thereby expanding protected areas within the park's boundaries.

In addition to its status as a Biosphere Reserve, in 1987 the Manu was designated a Natural World Heritage Site under another UNESCO program involving over 160 countries in a common commitment: safeguarding and caring for areas which, no matter where they are located, can count on permanent help in preserving their existence, beauty, natural and cultural wealth, their richness.

The title Natural World Heritage Site has reinforced even further the importance of the conservation of the Manu, in all aspects. There, where a traditional way of life still exists, where the birds surprise us with their colors, and where animals can be seen in the middle of the forest, lives a message for the planet about the importance of welcoming diversity.

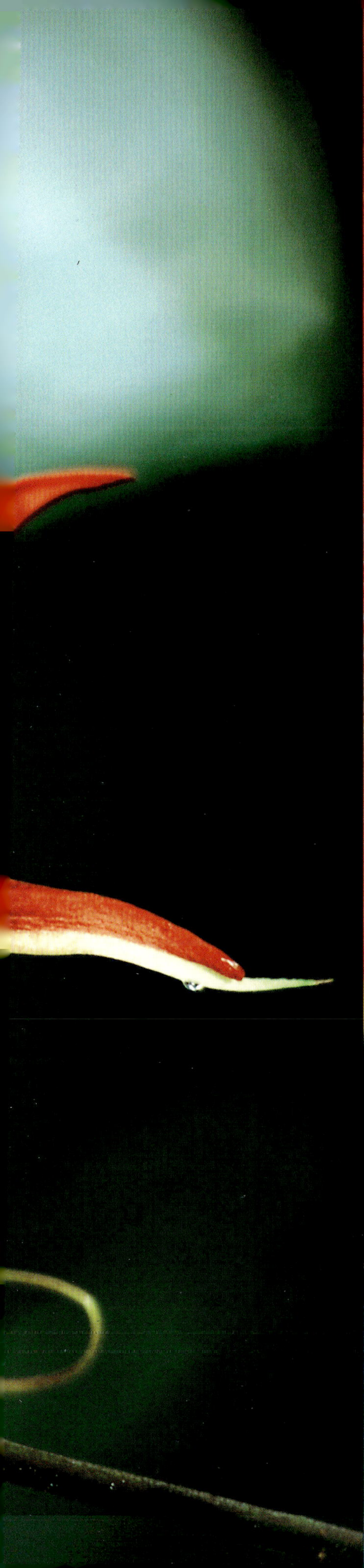

A young man from the Machiguenga people works as a guide in the park. Tourism has brought new prospects for the native population of the region, previously dependent on the cultivation of coffee for their livelihood.

During the rainy season, the water creates and recreates the forest: as well as ripping up huge trees and dragging them down its course, the river also spreads their seeds to other regions.

The cry of this small red howler monkey (*Alouatta seniculus*) can be heard for kilometers and enables contact to be maintained between groups.

The Amazon Forest is "stitched" together by a vast network of lianas. This one, known as a "monkey ladder", creates walkways suspended in the forest canopy. Thus, when a large tree dies or is cut, it falls, dragging others with it.

The river is the highway through the forest: all transportation is done in canoes or motorized river crafts that carry tourists, researchers, and natives to their destinations.

Environmental education among the inhabitants of the Manu involves especially the children. Species previously seen as food or pets are now protected by them.

The Manu National Park is home to several indigenous tribes that live in isolation, or have had minimal contact with the outside world. Using traditional techniques to hunt and fish, their lifestyle continues unchanged from that of their ancestors.

A group of caterpillars winds through the forest, as if it was one living organism. Together, they have a better chance of survival.

In Timpia, on the Machiguenga reservation located to the west of the park, the indigenous people coordinate a successful experiment in ecotourism.

Many Machiguengas live isolated lives in small groups in the forests of the Manu, speaking only their tribal language and cultivating the same lifestyle as that of their ancestors. For how long?

The rivers of the Amazon are the great breeding grounds of life. The waters of the Manu River, which flow through a good part of the park, are muddied due to the large amount of sediment washed down from the Andes.

The Amazon Forest tends to be monochromatic. Multicolored flowers therefore stand out, attracting insects and birds, as well as observers of fauna and flora.

In tropical forests, the scarlet macaw (*Ara macao*) is threatened with habitat loss and with predatory hunting for their colorful feathers.

Due to their protected status in the Manu National Park, red-and-green macaws (*Ara chloroptera*) can be seen in large groups on riverbanks, where they feed on clay. The minerals found in the clay neutralize the toxic effects of the seeds upon which the birds feed.

2 The many roads of the South.
Torres del Paine National Park.

Location Southern Chile.
Area 242,242 hectares.
Creation Set up in 1959 under another name and with only just over 4,000 hectares, in 1970 the park was renamed Torres Del Paine and in 1978 declared a Biosphere Reserve by UNESCO.
Ecosystem Paine has a varied landscape with much inter-linkage between ecological regions. In the lower plains are the Patagonian Steppes. On the mountain slopes is the Magellan Deciduous Forest and in the high altitudes, the Andean Desert.
Climate On the steppes, the climate is semi-arid and cold. The mountains are temperate with a mild summer and an extremely cold winter.
Relevant features Adventure and sports tourism (trekking, mountaineering, canoeing); development of projects with the private sector.

2 Torres del Paine National Park. The many roads of the South.

South America ends in a sort of point curving towards the Atlantic. It looks like a continental comma, the point of which ends in the cold and ferocious winds of Cape Horn, also the end for many sailors. Many call this point "the land at the end of the world", as it is one of the least inhabited areas on Earth. Living here is no easy thing. It is a gelid area, castigated by winds that can reach hurricane force. Everything seems extreme in the land at the end of the world.

Geological history and its tectonic movements have been responsible for this age-old scene of sound and fury. It all began around 70 million years ago when the Nazca oceanic plate was violently pushed under the South American continental plate, forcing up the Andes Mountains. In the southern cone area, where all had been plains, sediment and ice, the force and pressure were such that they forged huge blocks of granite in the deeper layers of the crust. In a final revolution, around 12 million years ago, the granite was extruded to form a mountain massif, one of the most beautiful in the world. This was the Torres del Paine, with its towers, horns, and mountain peaks.

Time, wind, and rain eroded the sedimentary rocks covering the massif, leaving an impressive formation of abrupt, vertical walls, some of them over 1,000 m high.

Paine means blue in the language of the Tehuelche Indians, and these impressive mountains are so-called perhaps because they appear to be part of the sky, or perhaps because seen from the plains of Patagonia – from where the Indians came – the granite takes on a blue cast. An intermediary formation between the eastern pampas (Atlantic Ocean side) and the high Andes to the west (Pacific Ocean side), the towers of Torres del Paine have created a climate specific to the area around them, temperate in summer and extremely cold in winter, but where year-round winds whistle up the slopes, enclosing the mountain in a gelid embrace.

It should be clearly understood that the Torres del Paine is not the Andes, and it is neither on the Atlantic nor on the Pacific. The massif has its own distinct identity, formed by a unique geology. Unbowed by the heights of the neighboring Andes, it reigns majestically over the surrounding pampas.

Set apart from the surrounding area, this kingdom has flora reminiscent of a garden, with vegetation alternating from low bushes and shrubs, whose flowers color the landscape in springtime, to coniferous woods and forests that, with the withdrawal of the ranchers and their cattle from the park, have regained their imposing stature. With the Paine National Park returning to nature, wild animals such as the guanaco *(Lama guanicoe),* a relative of the llama, the puma or mountain lion *(Felis concolor),* and the pampas fox *(Dusicyon gymnocercus)* have all bounced back from near-extinction. Now, twenty years later, the animals have built up significant populations and are relatively easy to be seen, particularly the guanaco and the fox. This recovery program has been aided by specially trained park rangers distributed around the eight sectors of the park.

The Torres del Paine is also covered by an almost surreal mantle of glaciers – rivers of ice that run down the mountains like great blue tongues, flowing into the lakes situated at lower latitudes. This is the western, frozen side of the Paine, with its high peaks and enormous snowfields forming a natural barrier separating Patagonia from the Pacific.

This ice-sheet, comparable only to that of Greenland and sectors of Antarctica, was formed at the end of the last Ice Age, around 10,000 to 12,000 years ago. Glaciers could also be classified as rivers, to the extent that they are water in motion. The ice descends from the peaks and higher altitudes extremely slowly, dragging along anything found in its way. As it reaches the lakes, already tightly compacted, large chunks are calved off to drift away as slowly melting icebergs. When an iceberg calves from the glacier, it falls into the lake with an enormous crashing sound, sending up huge waves.

If the mountains are monarchs, the mountain climbers are modern knights. The park and its mountaineering routes attract many expeditions and solo climbers from all over the world. They come to conquer the mountains and experience sensations to the limit. They come to meet the ice, the huge walls that appear as insurmountable obstacles, and to find themselves in the vastness of nature – this latter goal perhaps the most challenging of all.

It was around 1890 that the first English and German settlers decided to occupy the current area of the park, at first practicing subsistence agriculture, then cattle and goat ranching, causing a pernicious environmental impact, both on the vegetation and on the number of wild animals. By 1930, Chileans interested in the conservation of the wildlife proposed the creation of a park in the area. This became official in 1959, with the decreeing of the Lago Grey National Tourism Park, with an area of 4,332 hectares. In 1970, the area was significantly remodeled and enlarged to total 35,532 hectares and was baptized with its current name. In following years, the Paine was further enlarged to today's 242,242 hectares.

In 1978, UNESCO declared the area a Biosphere Reserve, further increasing international interest in the park.
The creation of the national park, along with its infrastructure during the 1980s, attracted a growing number of tourists. The first thing was to set up a circuit of trails leading around the towers, over a course of more than 200 km, allowing visitors to spend several days within the park.

In addition to well-defined trails, at each stage of the route, visitors can find some kind of accommodation, ranging from camping areas and refuges to hotels, some of them in the luxury category. The structure of the Paine National Park for receiving visitors sets an example for other national parks in Latin America, with options to suit every kind of tourist, from camping areas to simple mountain hostels, and even a sophisticated five-star hotel. An important aspect is that this infrastructure is restricted to the 4% of the park open to the public. The other 96% is off limits to visitors and is left virtually untouched by human interference.

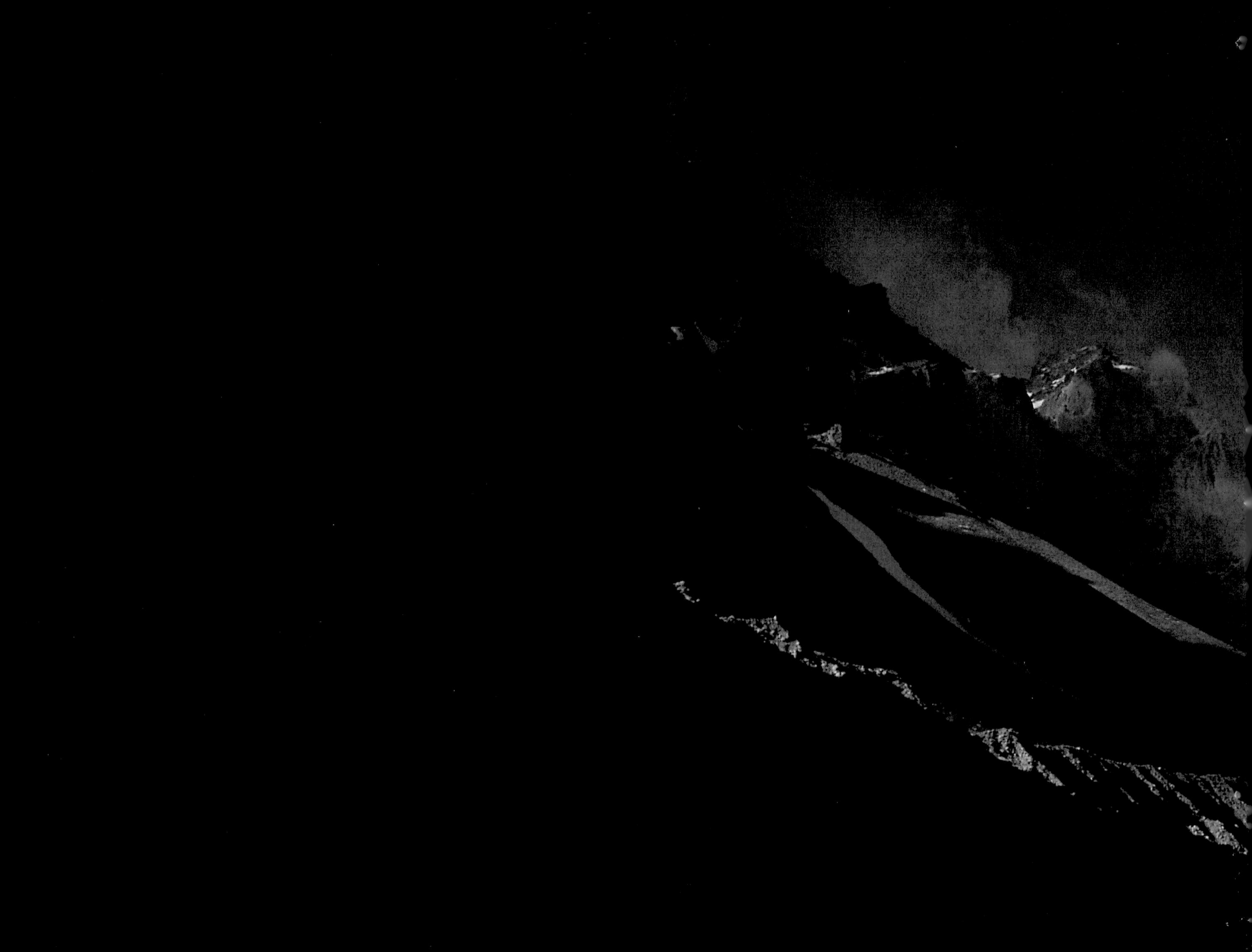

The conservation program of the park, allied with improvements to its basic infrastructure – its trails, safety structure, and the team of park rangers – has made Paine attractive to investment in tourism.

Establishing relationships with the private sector has been another example set by the Paine Park. Although ranging from accommodation and transport services to guide and equipment rent, many activities have been offered as concessions with contracts ensuring rigid quality control and many additional requirements. These services are continually monitored by park rangers.

On these visits, all commitments undertaken by contract are analyzed and the non-fulfillment of any of them carries with it heavy sanctions. In the same way, revenue passed on to the park is reviewed from time to time, so that the accord is always of mutual benefit to both parties. These assessments and reviews guarantee the quality needed to properly maintain the park. The agreements and concession leases are an important source of revenue for the Paine, increasing funds available for improvements needed in visitor control, in the maintenance of trails, and in the constant monitoring by park rangers.

The park has leveraged tourism throughout the region, creating a more profitable alternative to the previous economy based on cattle ranching, with its tendency to concentrate wealth and cause environmental damage. Tourism has also created jobs. Even in the small city of Puerto Natales, 150 km away and the obligatory port of entry to the park, tourism is now the main source of income.

The Torres del Paine National Park really does enjoy a privileged position, not only for its superb geology and landscapes, but also for the kind of commitments and relationships formed between the government and the private initiative. Favorable winds are taking the news to other national parks in Latin America: the frozen winds of the South.

The Paine massif resulted of an erosion process of thousands of years. Time, water, and the wind have washed away the sedimentary rocks, revealing today's impressive formations of vertical rock faces – some over 1,000 m high.

With the creation of the national park and the removal of the cattle, the number of guanacos (*Lama guanicoe*) began to increase, signaling a return to conditions prevailing before the arrival of Europeans.

The gradual global warming has deeply affected the world's glaciers and ice caps. In Torres del Paine, the glaciers are slowly retreating.

With the withdrawal of cattle from the area of the national park, the forest has begun to regain its former glory.

The Paine towers, "Torres del Paine", are the star attraction of the park. It takes a long and arduous trek to reach them. National parks usually reward visitors with a feeling of accomplishment.

The park is a mecca for mountaineers from all over the world, attracting numerous expeditions and solo climbers.

After the rigors of winter, fields of wildflowers transform the landscape.

This South America's unique ice field, comparable only to those of Greenland and Antarctica, was formed in the last Ice Age, which ended from 12,000 to 10,000 years ago.

Patagonia is notorious for its sharp changes in the weather. The winds can be so strong that the Brazilian blessing "may the wind blow to your benefit", becomes more relevant than ever.

3 The web of life in the Cerrado.
Chapada dos Veadeiros National Park.

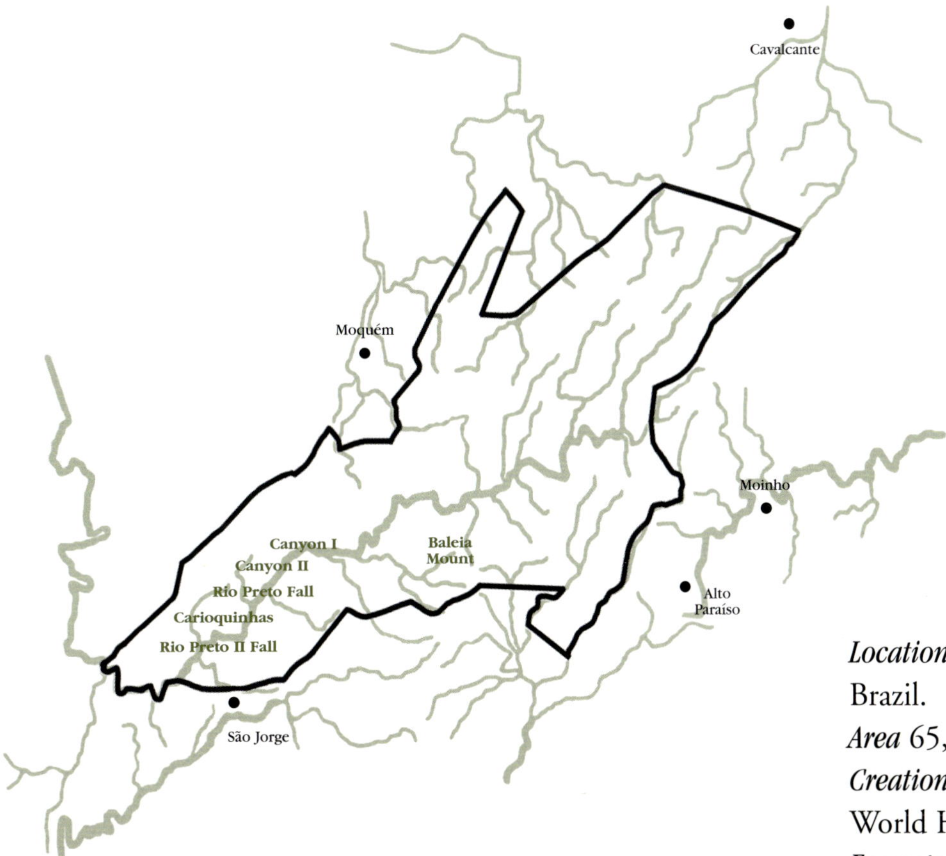

Location Northeast Goiás, central-western region of Brazil.
Area 65,514 hectares.
Creation Founded in 1961, it was declared a Natural World Heritage Site by UNESCO in 2001.
Ecosystem Highland Cerrado[1].
Climate At altitudes varying between 500m and 1,600m, the park has a semi-humid tropical climate, with a four to five month dry season and an average annual temperature of 25°C.
Relevant features Ecotourism and training of the local population as visitor guides. Private Natural Heritage Reserves (RPPNs) are located in the area surrounding the park, in addition to the Paranã-Pireneus ecological corridor.

3 Chapada dos Veadeiros National Park. The web of life in the Cerrado.

The landscape is impressive. Some of the canyons and plateaus of the Chapada dos Veadeiros look like fortresses, with their abrupt, vertical escarpments. Reinforcing the idea of isolation, caves and rivers at the bottom of the canyons remind us of unbridgeable moats.

Whether because of its broken geomorphology of ridges and scarps, or because central Brazil's "emptiness" really is pretty empty, the fact is that the disinherited from various regions have tried to make this part of the Cerrado their protected kingdoms. Some were successful, at least for a while.

The Avá-Canoeiro Indians of the Tupi linguistic family, for example, skilled at paddling their river dugouts, made the Chapada dos Veadeiros their refuge, almost to the last man. The object of systematic persecution from the earliest colonial times, in bitter fights with the Portuguese, in the 19th century, they decided to organize themselves into small groups easy to mobilize, but also easily forgotten. So much so that in the 1980s, when nobody remembered them any more, workers at a dam that was to flood the Serra da Mesa (west of the Chapada), along with officials from the Federal Indian Agency (FUNAI), found a group made up of a couple and two older women, who had for twenty years been hiding in the caves and mountains, fleeing every day, day after day, just running. There were no children among them, as Tuia, the younger woman, did not allow her pregnancies to come to term: babies would make too much noise and slow them down. After being contacted and then afforded protection in their own reservation on the fluvial island of Bananal, Tuia and her husband had two boys. FUNAI's hope is that some other small group will be found wandering the Chapada dos Veadeiros, and that among them will be girls, so that the once great Avá-Canoeiro nation does not become extinct.

It was also in the Chapada dos Veadeiros that groups of Afro-Brazilians, called the Kalunga, lived out their story of resistance. The records show that the Kalunga were Bantu, skilled in the arts of mining and metalworking, having been brought to the state of Minas Gerais in the 18th century as slaves. They escaped, heading further into the interior until they reached the Chapada, and there founded their dream of freedom.

They were able to do this thanks to the broken nature of the terrain, crossed as it was by rivers, scarps, and difficult trails. They managed to remain together in virtual isolation until the 1980s, when an astonished Brazil "discovered" the five Kalunga quilombo[2] communities, home to around 3,000 people.

Even revolutionaries sought refuge within the Chapada from 1925-27, during the Revolution of the Lieutenants. The young soldiers who wished to overthrow the government set off around the country under the leadership of Luiz Carlos Prestes, eventually finding sanctuary in the Chapada.

The Chapada dos Veadeiros consists of highland cerrado (its highest point is over 1,600 m), contrasting with the lower lands and plains of Brazil's Central Plateau.
Only 3% of the cerrado is considered highland cerrado, which combines a more temperate climate with the golden-green of natural pastures, enclosed here and there by abrupt escarpments and the vertical walls of tablelands and canyons. Close to watercourses – the Chapada dos Veadeiros has a wealth of rivers, including the headwaters of the great Tocantins River – groves of Buriti and other palm trees decorate the landscape with their waving fan-like leaves. The smaller trees with their twisted branches and thick bark, typical of this biome, explode into color after the beginning of the rains (October-November), contrasting with the permanently blue skies.

The cerrado, or Brazilian savanna, is the second largest biome in the country, covering one-quarter of Brazil, or around 1,700,000 km^2. It is an ecosystem extremely rich in species, a high proportion of which are endemic. Until the 1960s it was ignored (and therefore preserved), due to the idea that the soils were thin and unproductive. With the construction of Brasília, however, its plains began to be seen as suitable for extensive and highly mechanized agriculture. In the short space of 40 years, 80% of the ecosystem was destroyed.

Over much of the area, where once were golden grasses, stunted trees, and palms, today endless one-crop plantations spread to the horizon. It is likely that in the future, after the tractors and combine harvesters, there will come desertification.

The Chapada dos Veadeiros National Park reflects a little of the saga of the cerrado.

In 1961, soon after the foundation of Brasília, the Tocantins National Park was decreed by then President Juscelino Kubitschek. At the time, its area was 625,000 hectares. In the years succeeding the 1964 military coup, as had happened to the Indians, the quilombos and the lieutenants, the lands suffered an implacable persecution from ranchers and land speculators, who succeeded in reducing the natural vegetation to just over 10% of their original area. In 2001, after being declared a World Natural Heritage Site by UNESCO, the area of the park was enlarged to 235,970 hectares by the Brazilian government. The measure was, however, declared void by a court decision. A new project for its enlargement is currently underway.

Veadeiros was the name given to dogs that used to hunt the pampas deer (veado-campeiro, in Portuguese), previously abundant in this part of the cerrado, but today almost extinct.

Around 1940, the main economic activity in the Chapada was mining the huge deposits of rock crystal. In 1944, over 2,000 miners lived in the current area of the park. Years later, production of synthetic crystal for industrial use put an end to mining, leaving as witness a few areas of dug-up earth shining with thousands of tiny crystals. The presence of the rocks, considered by some to hold special energies, has attracted a different kind of business to the region. Mystics and others of an esoteric nature from all over Brazil and abroad come to the Chapada as a place for cure and meditation, giving a boost to ecotourism or "esotourism", as it is locally called. Nowadays, even after the discovery of the area by big-city tourists, mostly from Brasília, the mystical atmosphere is still maintained in the numerous post-hippie communities, temples, and even in a "flying saucer observatory", as the Isto Lookout is called.

The former miners and their families who stayed in the region's towns and villages, today face a totally new prospect in their lives: that of working with tourism and taking part in the daily life of the park as visitor guides. A special program set up by the Brazilian Environmental Agency (IBAMA), by NGOs, and by the Goiás State Secretariat for the Environment has been training personnel over the last ten years, creating a team of dedicated conservationists. The visitor guides earn their living from the park and, in addition to guiding tourists, are responsible for their safety, maintaining the trails, and controlling the established times, and restrictions. They are also part of the firefighting teams (fire is a chronic problem in the cerrado) and work in rescue and life-saving operations, for which they have received special training.

The tourist activity, combined with environmental conservation, has also spread to the surrounding landowners who, by taking advantage of statute No. 9,985 approving the National System of Conservation Units (SNUC), have made their lands into RPPNs, thereby receiving, in addition to exemption from rural taxes, support in developing ecotourism projects. These activities have proved to be far more profitable than raising livestock. Today, some of the most visited attractions in the Chapada are to be found outside the boundaries of the park.

And, of course, these RPPNs represent a considerable increase in the protected area, extending the corridors for animals, thereby ensuring a greater genetic pool for the region's species. The Cara Preta RPPN covers almost 1,000 hectares, and is one of the best examples, protecting as it does, the sources of the Preto River, responsible for most of the park's waterfalls.

In 2000, the Serra do Pouso Alto Area of Environmental Protection (APA) was created around the park, covering over 800,000 hectares, and extending over four municipalities.

With this tendency for growth of the protected areas, all of them aimed at preserving living relationships within the park, we can say that the Chapada dos Veadeiros is set on a new path, toward a new destiny. No longer closed on itself in isolation, like the inhabitants of times gone by, but rather stretching itself and reaching out toward other conservation units in the states of Goiás and Minas Gerais, it is now less of a fortress and more like a web, a web of life.

Notes

[1]Cerrado – Brazilian savanna

[2]Quilombo – communities of the descendants of escaped enslaved Africans

In springtime, the cerrado explodes in colors. Many of its flowers are well managed and gathered for sale, as is the case with the *Paepalanthus*.

Fauna in protected areas tends to be friendlier, due to feeling less threatened by the presence of humans – as with this toco toucan (*Ramphastos toco*).

The Cerrado ecosystem is one of the most varied in terms of life, but is also one of the most threatened of Brazil.

Several waterfalls on the Almácegas River are attractions in the area surrounding the park. The name of the river comes from the *cerrado* plant, whose balsam resin is used both medicinally and for its aroma.

By becoming a profitable activity, tourism has encouraged conservation in the Raizama region. Where previously there was a cattle ranch, now there is a Private Natural Heritage Reserve (RPPN), opened to public visitation.

Protection of the pristine waters of the park has always been a priority. The Private Reserve of Cara Preta (RPPN) was set up through private initiative, to protect the headwaters of the Preto River, the main river of the park.

For the watchful observer, the rocks in the riverbeds and surrounding area of the park are attractions in themselves.

After decades of being hunted, the jaguar's (*Panthera onca*) protection is now one of the greatest challenges for national parks.

The Cerrado, known as the Brazilian savanna, is the second largest biome in the country, extending over one quarter of Brazil's territory, around 1.7 million km^2.

4 Creativity in the living forest. Arenal Volcano National Park.

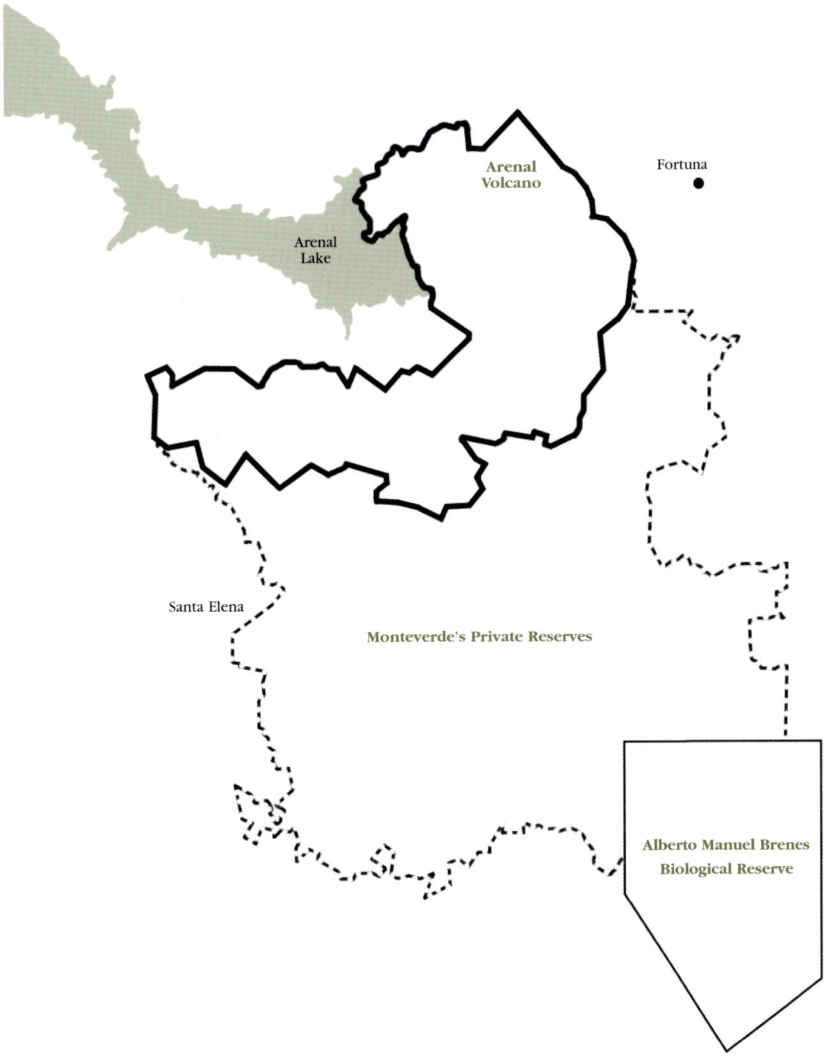

Location Tilarán Cordillera, northern Costa Rica.
Area 12,016 hectares.
Creation September 26, 1994.
Ecosystem Low-altitude zones have tropical forests, while mountain slopes are covered with cloud forest.
Climate The climate is predominantly tropical, with rainfall of up to 5,500 mm/year.
Relevant features Ecotourism in privately owned reserves, imaginative attractions, and connecting corridors for flora and fauna.

4 Arenal Volcano National Park.
Creativity in the living forest.

Costa Rica is part of a young land. Geologically speaking, it was born "yesterday" – just three million years ago, in a planetary history of 4.3 billion –, the result of the pressure of continental plates that raised the ocean floor, causing intense volcanic activity, which even today, vents its fury from time to time.

One such time was on July 29, 1968, in the northwest of Costa Rica, when everyone least expected it. The Earth spat heat and incandescent lava with such intensity that an area of 12 km^2 was devastated in a few hours, killing almost 100 people and destroying two villages in its way. The Arenal volcano, which people in the region simply call Mt. Arenal, opened craters that were never to close.

Costa Rica has a magnificent gift, typical of those who love life: that of seeing with new eyes and knowing how to transform. That is how it was with that "quiet mountain", today an active volcano. It fills the soul to look at its imposing shape on the horizon, at the pyrotechnic displays that occurr every night, and at the exuberant nature that surrounded it. It was then transformed into a park, establishing a direction and a vocation that was enthusiastically taken up by the whole country: that of environmental conservation.

Located in the north of the country, in the volcanic mountain range of Guanacaste, north of the Tilarán Cordillera and part of the San Carlos plain, the park sits on the shores of the large lake of the same name: Arenal. It is a rich region, with important watersheds, many of whose rivers flow into the lake, creating an enormous volume of water. This great quantity of water produces 40% of the country's electrical energy, and for this reason, the region is known as the "energy heart of Costa Rica". Since the quality and quantity of these waters are now fundamental to the country and the energy sector, the company responsible for the administration of electricity generation pays for the conservation of the environment. The beneficiaries are those responsible for the protection of the forests that ensure the water supply: the national park and the private protected areas, whether owned by NGOs or by individuals.

Until a few decades ago, the situation was very different. Costa Rica, as with every other developing tropical country, was chopping down its forests. Nevertheless, bit-by-bit, awareness grew that the conservation and sustainable use of the country's immense biodiversity were more valuable in economic and human terms.

In the nineteen seventies, a conservation program begun, involved the whole nation, and transformed Costa Rica into a country that earns revenue from keeping its forests intact, whether through ecotourism or from the rational exploitation of its rich biodiversity.

Costa Rica is a small country of just 51,000 km², but with an immense diversity of life, due to the meeting of species coming from both North and South America and by the variety of climates that occur, depending on altitude or proximity to the sea. Estimates put Costa Rica as having at least 4% of the planet's biodiversity, distributed in habitats that vary from near desert, to the humid cloud forests that occur on mountain slopes from sea level to 3,500 m.

Such a concentration of life has been prized by a succession of governments. They decided to establish a program to protect the remaining forests, reduced to little more than a quarter of Costa Rica's territory. In addition, on October 24, 1989, the National Institute for Biodiversity (INBIO) was set up with responsibility to do the inventory, the control of prospecting and of the announcement of the country's wealth of biodiversity. Multinational companies pay heavy royalties for the right to prospect for new active principles in nature, whether for medicinal or cosmetic use. INBIO is also responsible for the biodiversity inventory, and to do this, inhabitants of the areas surrounding reserves are trained as amateur taxonomists to collect species samples. Judging by the seriousness with which this program is being carried out, within a short time Costa Rica will have catalogued all the country's species.

In the Arenal Volcano National Park, another innovative initiative in support of the surrounding population has arisen in the Monteverde region, on the park's southern borders. There, specialized and imaginative ecotourism projects are creating jobs and opportunities.
These economic alternatives for the region's communities changed their inhabitants to be more interested in its environmental conservation.

Monteverde is home to a community that was traditionally linked to dairy farming and coffee plantations. However, stimulated by the benefits of conservation and the potential of ecotourism, the local population has set up a series of privately owned reserves, totaling over 25,000 hectares, where the visitor can find a variety of nature-oriented activities, in addition to several small hotels, guest houses, and other businesses providing infrastructure to local tourism.

It all started with Monteverde's 550-hectare Eternal Forest (Bosque Eterno), set up in 1950 by the region's ranchers, concerned about protecting the headwaters of Monteverde's water basin. In 1972, The Monteverde Cloud Forest Reserve was created by a group of environmentalists fearful of the rapid settlement of the region's forested areas. They began their work with the purchase of 328 hectares of land. A year later the reserve was passed on to the Tropical Science Center (TSC), a scientific institution working with various universities and which expanded the reserve to its current 10,500 hectares. A profitable system of trails was set up there, attracting mainly bird-watchers from around the world. Visitors are accompanied by specially trained guides. Over 400 bird species have been documented in the reserve, the star attraction being the resplendent quetzal (*Pharomachrus mocinno*) with its magnificent plumage, and found only in certain areas of Central America.

The Children's Eternal Rain Forest (Bosque Eterno de los Niños) has an even more touching story. It is a forest that has been saved by children. This dream began in 1987, in a small rural school in Sweden. At the time, the children were studying tropical forests and their rapid destruction. This situation so moved a teacher and a nine-year old boy, that, together with other students at the school, they started a fundraising campaign in the town to purchase threatened areas of forest. They chose the region surrounding the Arenal Volcano National Park, in faraway Costa Rica, to realize their dream of conservation.

They began with six hectares, but at the end of two years, they were able to acquire 7,258 hectares, through the success of the children's fundraising campaign. The initiative became famous around the world, and various other similar campaigns have occurred since. The reserve currently has over 17,000 hectares and other children's reserves have been acquired in other tropical countries. The International Children's Rain Forest Network (ICRN) has been set up – an NGO of teachers and students dedicated to protecting tropical rain forests.

In this way, even more private reserves have been created around the Arenal Volcano National Park, some just for conservation, others to take advantage of ecotourism. With the imaginative attractions that have arisen, the visitor is tempted to stay longer in the region biking, walking through the forest canopy on suspended walkways or climbing trees using mountaineering methods – tyrolese and rappel are two examples. In addition, botanical gardens, aviaries, butterfly enclosures, and orchid collections can be visited. All these attractions, together with the park itself, mean that the region is now visited by more than 50,000 tourists per year.

The Arenal Volcano National Park, together with the Monteverde reserves, form a biological corridor that links the park to the Alberto Manuel Brenes Biological Reserve, protecting 10% of the Tilarán Cordillera and 45% of the forests remaining in the region.

These conservation solutions are sponsoring life in all its forms, placing man and nature on the same side – the winning side. They are living proof that private conservation initiatives make great allies with national parks in creating new directions for the 21st century. The biological corridors are an illustration – extending and joining together natural areas and reinforcing our hopes for the preservation of a living and diverse world.

After erupting in 1968, leaving a trail of destruction around it, the Arenal Volcano became the star attraction of the park, with its constantly flowing lava and pyrotechnic shows.

In Monteverde, located in the area surrounding the park, various initiatives have made the forest even more attractive by allying conservation with ecological and adventure tourism.

With over 400 species, birdwatching is one of the region's main attractions. In the photo a keel billed toucan (*Ramphastos sulfuratus*)

In La Fortuna de Arenal, the community in the area surrounding the park administers several attractions such as this waterfall. Specialized and imaginative ecotourism projects create jobs and opportunities in the region.

The Monteverde region has attractions that encourage environmental education and awareness. Among them, living exhibits of butterflies, orchids, and frogs.

Costa Rica's enormous biodiversity has been highly prized by successive governments willing to protect the remaining forest, to create protected areas, and to provide incentives for private enterprises to do the same.

The red-eyed leaf frog's (*Agalychnis callidryas*) sensitivity to small alterations in the environment makes it an important indicator to the health of the forests in which it lives.

The Children's Eternal Rain Forest (Bosque Eterno de los Niños) started as a distant dream in a small rural school in Sweden. Twenty-two thousand hectares of forest are now protected through the Children's efforts.

In national parks, areas off-limits to tourism are designated to give maximum protection to both flora and fauna, where nature is left to its natural cycles, with a minimum of human interference.
In the photo: spider monkey (*Ateles geoffroyi*).

Cientists became aware of the forest canopy and its biodiversity details only in the last two decades. Nowadays, activities such as riding a tyrolese allow everyone easy and unforgettable access to the top of the forest.

5 A tradition of conservation allied with tourism. Nahuel Huapi National Park.

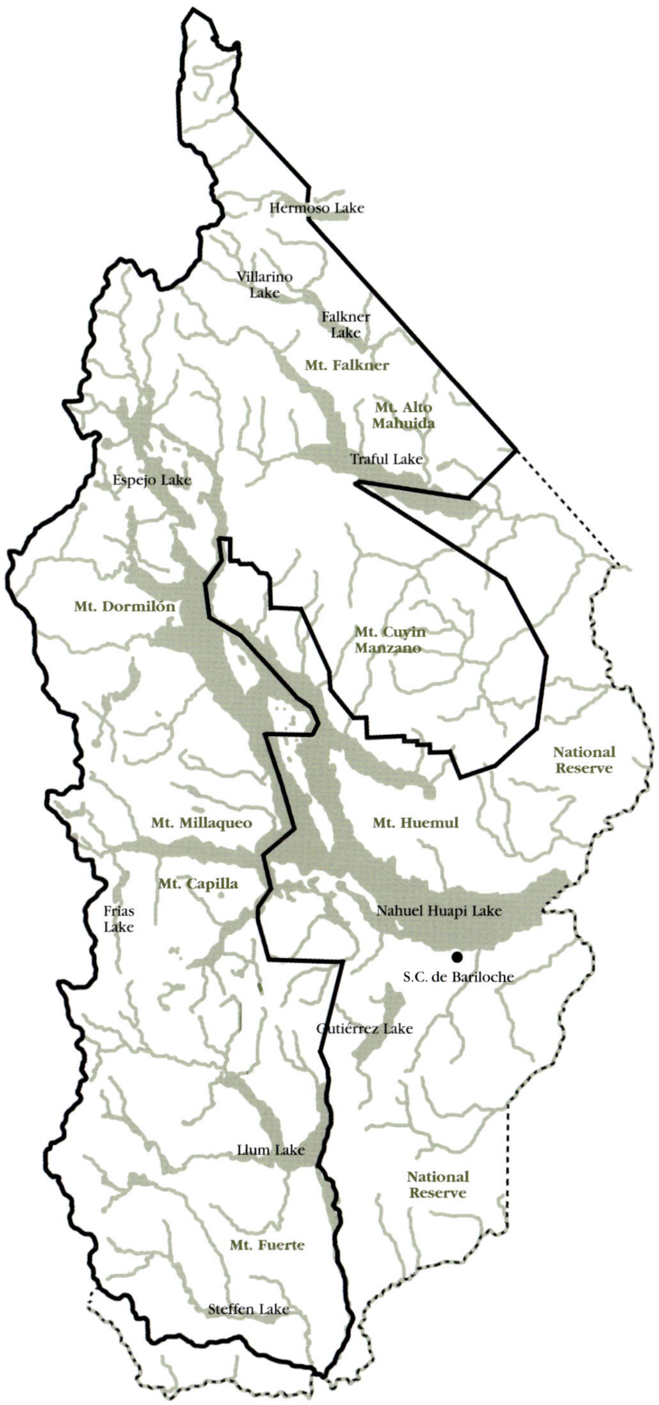

Location Northwest of Argentinean Patagonia, bordering Chile.
Area 428,000 hectares of national park and 330,000 hectares of national reserve, making a total 758,000 hectares.
Creation 1903, with the name Parque del Sur, covering 7,500 hectares. On October 9, 1934, the area was extended and the park received its current name.
Ecosystem Differences in relief and climate form four ecological life-zones in the park: the high-Andean, the humid forests, the transitional, and the dry steppes.
Climate Temperate and cold, with well-defined seasons and dry summers, with average temperatures between 14°C and 16°C. Wet winters, with snow and average temperatures between 2°C and 4°C. Precipitation varies from 4,000 mm/year in the forest zone to just 600 mm/year in the steppes.
Relevant features Ecotourism as a solution to conservation, to occupation of the land and as an economic alternative; well-trained park rangers and a volunteer force, with infrastructure provided through concessions tendered to the private sector.

5 Nahuel Huapi National Park. A tradition of conservation allied with tourism.

The discovery of new South American worlds by Europeans at the beginning of the 16th century was coupled with a great hope to find an earthly paradise. Together with the caravels and the men that crossed the ocean came dreams of finding civilizations or cities that would be extraordinary for both their material and human richness. The region of Argentinean Patagonia was also part of these dreams, enveloped by the mists of myth. According to reports from colonial times, the place was thought to hide a "city of Caesars" or an "enchanted city" of immense wealth where all men were equal. Many departed on expeditions that originated in lands that today belong to Chile.

On the Argentinean side, however, up until the end of the 19th century, this part of Patagonia belonged only to the native Indians, wild animals, and the wind on the waters of the great lake. This was until the arrival of Francisco Perito Moreno, an anthropologist and paleontologist from Buenos Aires, with his obstinate quest to find the true wealth of the region: its nature and the life of the native tribes. Moreno arrived in the region in 1876 and hoisted the Argentinean flag on the shores of Lake Nahuel Huapi, an immense lake with glacial blue water. From there he headed south and named four more lakes found on this route, as well as an impressive mountain, Mount Fitz Roy. Moreno was so dedicated to exploring the region that, years later, he would be appointed by the Argentinean government to defend the frontiers of his country in a dispute with Chile. The neighboring country claimed a good part of these lands, and wanted the border to be set at the limits of the headwaters of the streams flowing into the Pacific, most important of which was the Manso River. Moreno was able to convince the British arbitrators to take the summits of the highest peaks as the dividing line. As a result, Argentina annexed an enormous area to its territory, being one of the most beautiful in South America.

In 1903, intended on protecting the scenery and its native fauna, Perito Moreno, by now a national hero, donated to the government three square leagues (7,500 hectares) on the shores of the lake, on the condition that they would be preserved as a "public natural park", for the use of present and future generations. Thus was born the first national park in Latin America, the Parque del Sur, or Southern Park, the third to be proclaimed in the Americas, after Yellowstone in the United States in 1872 and Banff in Canada in 1885.

In 1907, with a new policy of giving prominence to national parks, the Argentinean government extended the park to cover 43,000 hectares. In 1934, with the creation of the administration, today known as the National Parks Administration (APN), it became known as the Nahuel Huapi National Park, with an area of 428,000 hectares. Alongside the national park, another great area was protected as the Nahuel Huapi National Reserve, with more than 330,000 hectares, acting as a buffer zone. A total of 758,000 hectares are now protected.

The name of the park came from the Araucanian native language – *nahuel* (jaguar) and *huapi* (island), and can be translated as "island of the jaguar", an allusion to today's Victoria Island in Lake Nahuel Huapi. *Nahuel*, the jaguar (*Panthera onca*), a species not documented in the region, is thought to have come from the totem of a family of the Puelche tribe who inhabited the island; or else, according to another version, the name originated with the bravery of these Indians, who had accordingly been called jaguars.

At the time, even after the borders had been secured by Moreno, the government of Argentina was still fearful of counterclaims over this beautiful and sparsely inhabited region. Therefore, instead of military reinforcements, possibly provoking claims in even more remote regions, the government decided on the more intelligent and peaceful strategy of occupying it through the means of conservation and tourism.

Years later, in 1947, the government initiated a series of partnerships with private capital, the first of which was a concession: that of the famous Hotel Llao Llao (pronounced "djow djow"). The policy of tendering service concessions within the park is nowadays a model to be followed, with the private sector and towns surrounding the park making their living from businesses related to tourism, which then give back some of their profits to benefit the park. An example is the concession for the Hotel Blest, on the shores of Lake Nahuel Huapi. The concessionaire has, in addition to the hotel, built a well-equipped visitor center for the park.

The city of Bariloche, with an area of 28,000 hectares, had around 110,000 inhabitants in 2002, living mainly on tourism leveraged by the national park. Bariloche is today an oasis of prosperity within Argentina.

The flow of visitors suffers no ill effects with the changing seasons. On the contrary, weather variations create new landscapes and different activities. In winter, the snow-capped Andes attract skiers and other winter sports enthusiasts. During summer, the considerable number of visitors go hiking on the many trails, enjoy canoeing in the rivers and sailing on Lake Nahuel Huapi, an immense body of water covering 560 km^2 and with depths of up to 454 meters at some points.

The diversity of beautiful scenery is an invitation to go hiking. There are basically four ecological life-zones in the park, each highly distinctive. The Andean highlands above 1,600 m have sparse vegetation and are covered by snow in winter. The mountains are home to such spectacular animals as condors (*Vultur gryphus*) and the huemul (*Hipocamelus bisulcus*), the Andean deer, which seeks shelter here in summer. The park also protects equally significant remnants of Patagonian Andean forest, including woods of arrayanes (*Myrceugenella apiculata*), a local species of araucaria pine (*Araucaria araucana*) and a local oak species (*Nothofagus obliqua*). It also protects a temperate rain forest, called "selva valdiviana", containing significant biological diversity, and which, in addition to temperate climate species, such as the remarcable coihue (*Nothofagus betuloides*), is home to various species typical of forests of warmer climates, such as liana vines, ferns, bamboo and even a species of parakeet (*Enicognatus ferruginea*), which give a tropical feeling to the forest. These forests provide food and shelter to numerous animal species. The steppes, with their low rainfall and grassy vegetation, are the last of the life-zones giving the Nahuel Huapi National Park its great diversity of climate, vegetation, and scenery.

The park is interconnected with other reserves in both Argentina and Chile, forming a complex of protected areas totaling over 1,600,000 hectares. It is an important ecological corridor, guaranteeing genetic exchange and the diversity of many species.

Taking care of all this complex web of relationships are the Nahuel Huapi Park rangers, a cohesive group of professionals, trained on worldwide standards and who carry out quality environmental conservation in Argentina. These are the real guardians of the park, with long experience in the region and trained at the National School of Park Rangers, the first in Latin America, which, until recently was based on Victoria Island, in the middle of Lake Nahuel Huapi.

Courses are subsidized by the government, last three years and are at the technical university level. Candidates must undergo a strict selection process. This course is usually open only to those who have already worked as volunteers in a park and is offered by the National Parks Association (APN).

This program for volunteers – aimed at young people from regions in which there are national parks or nature reserves – also provides conservation education in the communities, resulting in a commitment to the environment in the future, not to mention a career plan, as many young people who work as volunteers in national parks end up becoming park rangers, a highly appreciated profession in Argentina.

Today, in Nahuel Huapi, park rangers face the challenge of reconciling the growing demands of tourism and immigration with environmental equilibrium. The method found has been to integrate the community into the process, including decision-making over the park management. A consultative council has been created, the park ticket sales has been outsourced to two philanthropic institutions, with part of the revenue allocated to social programs in Bariloche, and the operation of a system of six mountain refuges has been leased to the local Andean Club. The park is also forming partnerships with local governments, technical institutions, and NGOs to offer environmental education programs in local schools. Extension courses are being given to small-scale farmers in the surrounding area, so that they too can benefit from tourism, and earn their living in a way that causes less environmental impact than, say, ranching.

Another noteworthy program in Nahuel Huapi is that of preventing and combating fires, a common and devastating occurrence in the cold, dry climate of Patagonia, with its scrub vegetation. In Argentina, every park has its own trained fire fighting force. Since the Nahuel Huapi National Park is in one of the country's areas of greatest risk, and which holds the area's largest concentration of inhabitants, the Argentinean Firefighters Center is based in Bariloche. Over fifty firefighters are always on call – some of them doubling as park rangers – supported by helicopters especially adapted to douse fires, and by aircrafts to transport personnel.

Fire is one of the greatest threats to national parks all over the world, and in Argentina, this is combated through prevention, efficient communication, and a solid infrastructure.

Nahuel Huapi Lake, the heart of the national park – formed by a retreating glacier – covers 557 km^2 and has a maximum depth of 454 m.

Nahuel Huapi National Park receives around one million visitors per year.

Lakes abound. Twenty percent of the national park is covered by water.

Several waterfalls formed by snowmelt can be seen at the foot of Mount Tronador. Glacier, snow, and water mix in a veritable lesson of natural history. The special geography of the park can be clearly seen in the formation of its great lakes.

At an altitude of 1,600 m, the Lopez refuge is reached after a long and arduous trek. Here, beauty is all around: views of the lake, the evening lights of Bariloche, and the night stars.

Part of the Puerto Blest region, the wet temperate rain forest of Valdiviana was donated to Argentina by Perito Moreno, on the condition that it would become a protected area, thereby becoming Latin America's first national park.

The Berghof refuge, where Otto Meiling, father of Argentinean mountaineering, spent much of his time, has been preserved and made into a refuge under a leasing arrangement.

The first Latin America school of park rangers was set up in Nahuel Huapi. Park rangers are indispensable to the conservation of national parks.

The park's attractions are suited to all tastes and seasons, and include skiing, trekking, photographic tours, canoeing, climbing, and horseback riding, among others.

The daily spectacle of colors and cloud formations in Patagonia is a result of constant winds and the purity of the air. In the photo: Cerro Cathedral in profile.

6 Paradise Rebuilt. Tijuca National Park.

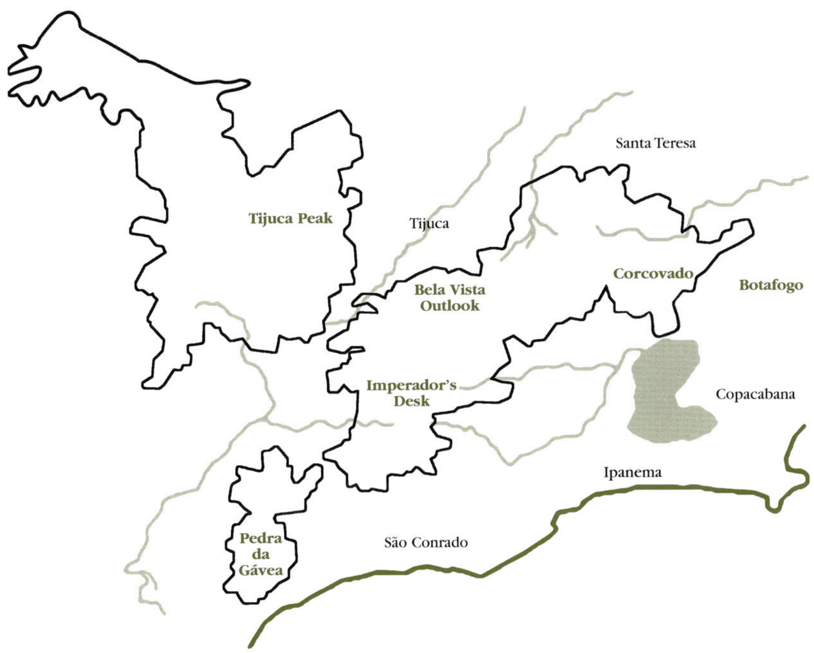

Location Rio de Janeiro (RJ-Brazil).
Area Currently 3,953 hectares after a 20% expansion in 2004 in the face of surrounding urbanization pressures.
Creation 1961. Despite having been made a public area of environmental protection in 1861, it only became a National Park one hundred years later. In 1991, it became part of UNESCO's worldwide network of Biosphere Reserves, as part of Brazil's Atlantic Rain Forest Biosphere.
Ecosystem Essentially atlantic rain forest, one of Brazil's areas of greatest bio-diversity, as well as the most threatened.
Climate Tropical: high temperatures and high rainfall (sometimes in excess of 3,000 mm per year).
Relevant features Leisure, trails, and reconciliation of conflicting interests.
This immense green area crowning the city of Rio de Janeiro is known as the Tijuca Forest by the people using its more than 90 trails. A historic model for reforestation in degraded areas, today the park is also a notable example of successful shared management between the federal and city governments, together with a wide public input through the Conservation Unit's Managing Council.

6 Tijuca National Park. Paradise Rebuilt.

It was said to be paradise on Earth. The Florentine Américo Vespúcio thus described the site of modern Rio de Janeiro in 1502, on his second visit to Brazil. The affirmation appeared in his third letter, which was later known as *Mundus Novus*, and became a bestseller in Europe, translated into numerous languages. Yes, Rio de Janeiro was paradise on Earth, due to its geographical beauty, luxuriant vegetation, and the natives – mainly the females – who had none of the European reserve of the time.

After Vespúcio's letters, the Tupinambá Indians, as well as the natural beauty, were left in relative peace for almost half a century until the French arrived in 1555. More than 600 men were led by Mr. Villegagnon with the firm intent of establishing Antarctic France, as it was to be called, in these lands.

The French were interested mainly in barter of artifacts, tools, and European trinkets for whatever the forest and the Indians could offer. And what was mainly on offer was "pau-brasil" – Brazilian rosewood – from which an extracted dye would color Europe and the clothes of the nobility of a deep, fleshy red. They took so many tons of wood and complained so much of the difficulties of the Atlantic crossing that a Tupinambá chieftain lost patience with so much greed:

" (...) You white people are crazy, since you cross the sea and suffer great hardship, as you said when you arrived, and work so hard to accumulate goods for your children and for those who will survive you! Will not the land that feeds you be sufficient to feed them also? We have fathers, mothers, and children that we love; but we are certain that after our death, the land that nurtured us will nurture them as well and for this reason we are able to rest, with no further cares (...)"

The Europeans, however, did not know how to rest. Motivated by a growing mercantilism, they wished to accumulate the maximum, without thoughts on the effects such an attitude could have on nature.

The Portuguese, who settled more to the south and judged themselves the rightful owners of Brazil (once again forgetting the ancestral rights of the Indians) were also upset at the presence of their European neighbors. Calling the French "brasileiros" – those that stole "pau-brasil" –, giving the word the connotation of pirates and smugglers, they created an expression that baptized an entire nation. And they did not rest until they had killed or expelled all the French and their Indian allies. The war lasted twelve years. Over 60,000 French and Indians died. Concomitant with the blood bath, they founded the city of São Sebastião of Rio de Janeiro, in that earthly paradise, in 1567.

It is impossible not to be uneasy imagining the symbolism of the events of the 16th century: a nation that gained its name from contravention, a country that was founded along with the destruction of nature, and a city that was founded upon a massacre. The Tijuca National Park is thus a parable about the power of freewill. The inhabitants refused to give in to destiny and they rebuilt nature, joining forces and choosing cooperation as a form of management.

However, the fate delineated in the 16th century still tried to impose itself. The French were driven out, but the curse that surrounds all paradises soon began: the fall of trees. The Portuguese began burning the forest in earnest to clear land for pasture and set up sugar plantations. In the 18th century, fueled by the boom in gold mining, Rio became the supplier of food and animals for the mines of Minas Gerais.

Even so, some green remained on the higher ridges, at least for a while. In 1808, with the arrival of the Portuguese Royal Family and their court, and with the consequent opening up of the ports to friendly countries, recently arrived foreigners headed for the still-forested upper slopes of Rio de Janeiro to plant coffee. Economic interest once again predominated. In around four decades, the devastation of Tijuca was so great that the city started to experience a terrible scarcity of water due to the silting up and lack of drinkable water in the rivers, mainly in the Carioca and Paineiras, which supplied the city. The government was forced to do something about the serious problem caused by deforestation. But how were they to expropriate and reforest lands that were so valuable and productive?

The popular saying that every cloud has a silver lining certainly rang true in 1843, when the "borboletinha" pest devastated the coffee plantations on Rio's hills, causing land prices to plummet. The government was then able to expropriate the owners. This was the embryonic birth of Tijuca National Forest.

Hired by Emperor Pedro II, Major Manoel Gomes Archer of the National Guard, with the help of a few slaves and several waged workers, was able to plant over 80,000 trees in 12 years, most of them typical of the Atlantic Rain Forest. The succeeding administration planted a further 35,000 seedlings, in addition to beautifying the forest with infrastructure for leisure activities. The forest was reborn and was given the status of national forest by the government in 1861.

Nature then followed its own course, or as the Major said, "The success of reforestation will depend much more on the action of time rather than on the actions of men."

In 1500, the atlantic rain forest, vegetation typical of the entire Brazilian coastline, covered over one million km^2. Over succeeding centuries, it was rapidly cut down by economic boom-bust cycles and by the expansion of Brazil's largest cities, as they advanced into the surrounding forest.

What saved the atlantic rain forest (today, only 7% of this Brazilian biome survives) was the geography of the Serra do Mar coastal range. The genetic banks are restricted to areas of difficult access, such as mountain tops, ridges, and deep ravines.

It was from the top of another range, Paineiras, that Major Archer took seedlings and created a nursery from which the future national park would be replanted.

In 1973, over a century on from Archer, Biologist Adelmar Coimbra Filho and Professor Aldrighi, chief of the park at the time, started an extensive program to reintroduce animals and birds. Today, Tijuca's fauna number over 100 species.

At the end of the 1980s, with the creation of IBAMA, in 1989, with the promulgation of a new constitution in 1988, giving increased priority to environmental questions, and finally, with the UN Conference on the Environment (RIO-92), held in Rio in 1992, the Tijuca National Park saw itself faced with new and greater challenges. Examples are the safeguarding of both conservation and leisure activities within its boundaries, finding a solution to the lack of funding it received as a federal Conservation Unit versus sustaining the huge local demands for its use; and reconciling often divergent interests between a federal administration and its urban location.

Conflicts of interests that appeared to be difficult were once more creatively resolved. A shared management was set up for the Tijuca National Park between IBAMA and the city government of Rio de Janeiro, each with functions set up in the partnership. One could say that this was perhaps inspired by the natural organization of the atlantic forest which, contrary to the temperate forests of colder countries, survives more through cooperation than competition.

Many seeds were planted as a result of this union. IBAMA, due to its expertise in the administration of Conservation Units, was responsible for the revenue generated by the park and for its investment in everything related to environmental improvements such as: personnel training, research projects, maintenance, management tools, as well as new strategies and partnerships between NGOs and the government. IBAMA is the bridge that links the park with the government and with worldwide environmentalism.

Rio's city government takes care of all that is related to the park's involvement with the city, its links to the State Government Environmental Secretariat and all that this entails, and all infrastructure for receiving the public. This ranges from maintaining trails to managing the Environmental Defense Groups and the Tourist Support Group. In addition, it consults with people living on the borders of the park and the general public to create projects aimed at involvement and protection.

The park will continue to innovate and set an example, if the just fight of Sônia Peixoto (the park's Director linked to IBAMA) for environmental assessement is heeded. The clean water supplying the city, the use of the land by transmission lines and communications antennae can, and should be, remunerated.

With all the improvements, the Tijuca National Park has again become popular with the city's inhabitants and with tourists from around the world. Visitors currently number 1.5 million per year. Most of them are blissfully unaware that a little more than one hundred years ago, the scene was one of environmental devastation, and that human hands caused nature to stage a comeback; living proof that, with goodwill, human beings can act as God and rebuild. After all, the seeds of our right to paradise lie within each one of us.

Symbol of Rio de Janeiro and one of the national park's main attractions, Christ, the Redeemer, bestows his mountaintop blessing on the atlantic rain forest and the city below.

The Tijuca National Park is one of the world's largest urban forests. The result of successful reforestation, it is where the locals go to breath, bath in a waterfall, and relax in the shade.

The national park also protects springs that supply fresh water to the city of Rio de Janeiro. The possibility of some form of environmental compensation for this protection is currently under consideration.

Native fauna can be found within the Tijuca National Park, such as this porcupine (*Coendou prehensilis*), which finds solitude in the treetops.

Rio de Janeiro and Guanabara Bay viewed from the upper part of the park.

In the 19th century, a process of recuperation was begun, with the planting of over 115,000 seedlings of native species. One hundred and forty years later, the forest has regained its equilibrium.

The forest gradually rejuvenates itself. "The success of reforestation will depend much more on the action of time rather than on the actions of men," said Major Archer, back in the 19th century.

A detail of flora native to the atlantic rain forest. Due to it being one of the planet's most diverse ecosystems, and at the same time highly threatened (only 7% remains), it is considered today as one of the world's "hot spots".

Every year, over 1.5 million people visit the Tijuca National Park for leisure or to practice a sport surrounded by nature.

The Tijuca National Park is surrounded by an incessantly growing city, hence the importance of policies that involve the surrounding population.

7 Conservation without borders. La Amistad International Park.

Location Talamanca Cordillera, southeastern Costa Rica and northwestern Panama.
Area 400,929 hectares (193,929 hectares in Costa Rica and 207,000 in Panama).
Creation February 4, 1982 in Costa Rica; September 2, 1988 in Panama. Declared a Biosphere Reserve in 1982, Natural World Heritage Site in 1983 (La Amistad Costa Rica) and in 1990 (La Amistad Panama).
Ecosystem Tropical forests at low altitudes, cloud forest on the mountain slopes.
Climate There is great climatic variation depending on the altitude, but predominantly tropical with high rainfall (5,500 mm/year).
Relevant features A model for international cooperation and the promotion of peace, ecotourism and adventure tourism.

7 La Amistad International Park. Conservation without borders.

The forest has no borders. It just spreads. Not that it spreads wherever it wants, setting its own limits. The forest only grows where it is wanted, where the geography embraces and welcomes it, letting it expand with no limits to its self-expression. It spreads where the climate invites it. The forest does not lead. It is led by the rivers that disperse its seeds, and by the animals that feed off it in one place, taking its seeds to germinate in another, further ahead.

The animals do not respect borders either. They wander in search of the best food and the best partners for breeding and exchanging genes. For the animals, a life with limits is no life at all.

Many of the people of the forest, at least in their origins, also ignore borders. They always look for renewal, leaving today's lands to rest, to return tomorrow, and then move on again, and so on, to where new life beckons.
Life obeys no limits. It is always moving ahead, naturally.

It was, therefore, a natural idea to think of a park that would cross borders, where all life could continue its course uninterrupted. The proposal was put into practice by Costa Rica and Panama in 1982, with the signing of an agreement creating La Amistad International Park (PILA), the friendship park. Various countries in Central America were concerned about the environmental degradation that surrounded the rich Talamanca Cordillera, a mountainous area that is home to seven of the twelve ecological regions typical of Central America, in addition to representing an important area for genetic exchange for the fauna and flora. It is the central point in the Meso-American Biological Corridor.

This corridor, which could well be called a highway, due to its length, is in fact a highway of life, a dream of uniting the last North American forests with South America's Amazon.
The Meso-American Biological Corridor, and its implementation, is a program that plans to extend the La Amistad International park's example of international cooperation to a further six countries, from Mexico to Panama, encompassing almost 40 protected areas and linking them through biological corridors.

It is certain that there has always been a historical will for union in the region, even in terms of the geology. It began three million years ago, when the shape of the world's landmasses was slightly different from that of today. There were two separate Americas – North and South. However, the movement of the continental plates caused intense volcanic activity, gradually resulting in the formation of a mountainous land bridge, a route for communication and exchange, today called Central America. North and South America were thus united. Even today, the region has active volcanoes, such as Arenal in Costa Rica.

For millions of years, life has been moving over this land bridge. North American species moved down, South American species moved up, causing one of the world's greatest concentrations of biodiversity at this meeting place, as can be seen in the Talamanca Cordillera. There are at least 115 species of fish, 250 species of reptiles, 215 species of mammals, and 560 bird species. Estimates show La Amistad as having 4% of the world's biodiversity.

The 400,929 hectares of the la Amistad International Park protect a large part of this cordillera, encompassing the largest tropical forest in Central America, containing hundreds of endemic species, those that only occur in this region. It is one of the last significant refuges in Central America for species threatened with extinction. Birds such as the resplendent quetzal (*Pharomachrus mocinno*), with its emerald-green and red plumage, and the imposing harpy eagle (*Harpia harpyja*), the national bird of Panama, that reaches over 1 m high and a wingspan of almost 2 m, are just a few examples in the region, a mecca for bird watchers. Among the mammals, the park is home to various groups of wildcats, among them the spotted jaguar (*Panthera onca*), as well as endemic and threatened species such as Baird's tapir (*Tapirus bairdii*) and the squirrel monkey of Costa Rica (*Saimiri oerstedii*).

Because of the extensive altitudinal range – 100 m above sea level to over 3,500 m in the mountains – the flora of the park is estimated to be the most diverse in the world, when compared to other reserves of a similar size. Over 10,000 plant species have been counted, in addition to 900 species of lichen and over 1,000 fern species. Endemism is estimated at 30%. The vegetation changes with the altitude, as do the ecological life-zones. Tropical humid forest stretches from sea level to 800 m, with high rainfall and giant trees, such as the sumaúma (*Ceiba pectandra*), that grows to 55 m. A little higher up, climbing to 3,000 m, is the premontane and lower montane vegetation, that covers most of the park, whose most common trees are mountain cypress (*Podocarpus macrostachyus*) and several species of oak (*Quercus sp.*), growing to 40 m tall. Above 3,200 m is subalpine páramo (*moorland*), characterized by shrub vegetation, and partially dominated by a species of dwarf bamboo (*Chusquea subtessellat*).

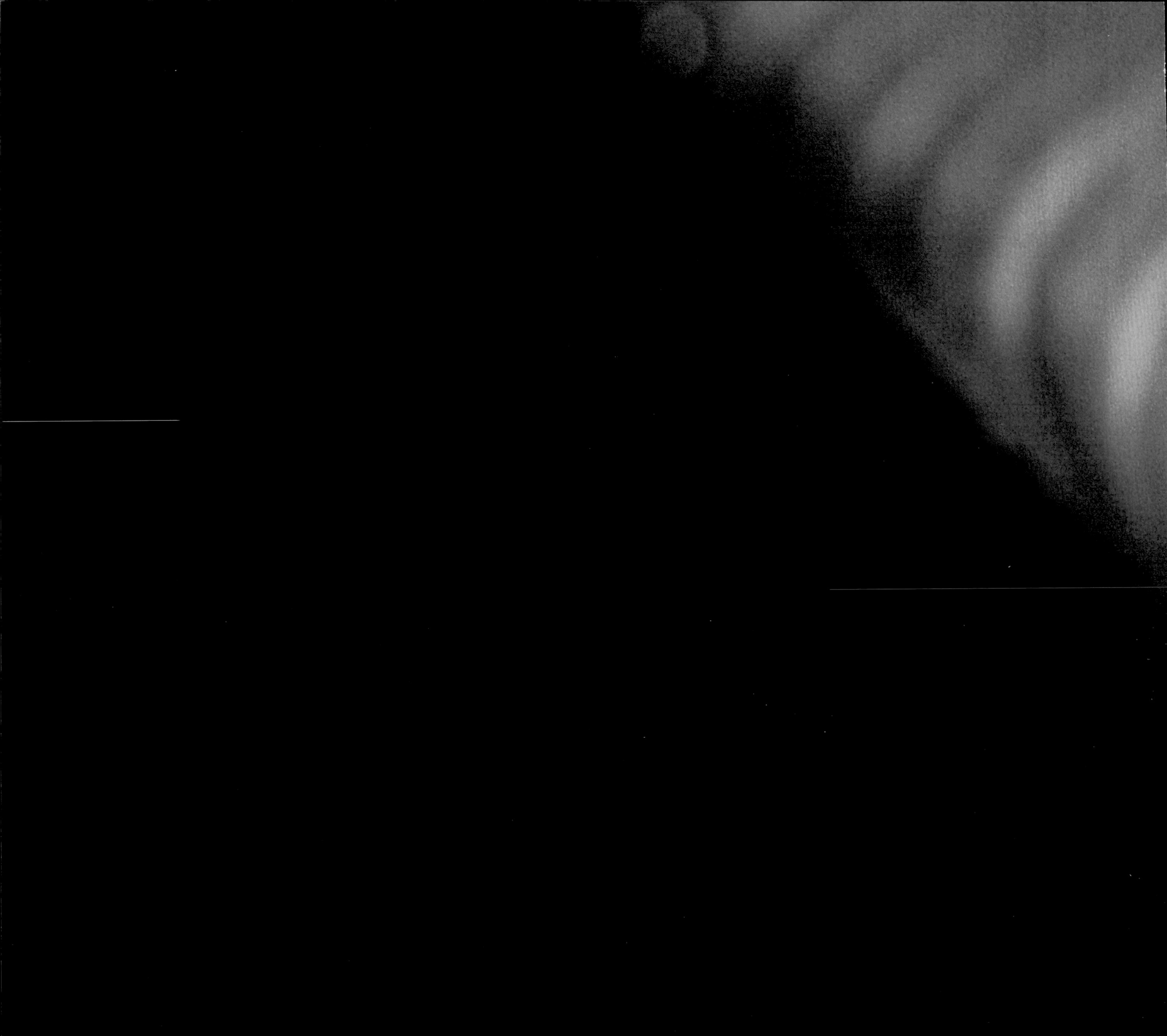

The topography of the region is the result of its volcanic past. Broken relief is topped by highlands that frequently rise over 3,000 m, and divided by valleys carved by innumerable rivers. The Talamanca Cordillera is an important watershed for the two countries in which it is located. However, it is not only pure water they have to share. The park has created opportunities for joint action in areas such as fire control, environmental surveillance, and the exchange of scientific data and research.
The two countries and their administrations aim at acting in synergy in the park, thereby economizing resources for other projects. Instead of the tension normally found in border areas, the park suggests a readiness for cooperation, with money spent on conservation rather than on weapons.

Various indigenous areas surround the international park. Among the cooperative agreements made since its creation, is the safeguarding of free passage for the indigenous population among their lands, no matter from which country they have come, nor to which country they are going. La Amistad International Park is an example of tolerance and flexibility overcoming borders.

The work of La Amistad International Park in the conservation of the Talamanca Cordillera, and the efforts of the governments of Costa Rica and of Panama in establishing and broadening integration and cooperation, led, in 1985, to the recognition of the area as a trans-national World Natural Heritage Site, encompassing the entire international park, in addition to six surrounding protected areas, making a total area of around 570,000 hectares. The inclusion of a region in the World Heritage List means that the international community recognizes that area as having enormous value to present and future generations. A title such as this, awarded by the United Nations Education Science and Culture Organization (UNESCO), benefits the area even more, by recognizing its importance, attracting international tourism and enabling the opening of lines of credit for the implementation of infrastructure.

In 1982, the park and other conservation units in the region, including some indigenous areas, were declared a Biosphere Reserve, as part of UNESCO's MaB Program. This double international recognition, encompassing various conservation units and indigenous areas, totals over 900,000 hectares.

La Amistad has been another example of peace and union demonstrated by Costa Rica, a country which opted for abolishing its armed forces in 1948, and which has allocated the resources saved to health and education programs ever since. In the seventies, the country used funds obtained from the debt for nature swap – whereby foreign debt was written off in exchange for a commitment to conserving nature – to set aside a large part of the 25% of original forest still remaining in the country as conservation units. A fuel tax was also imposed and used to fund conservation and environmental recuperation of privately owned areas.

Another successful initiative was the creation of the National Institute for Biodiversity (INBIO) in 1989, responsible for compiling an inventory of the country's biodiversity. The results are compiled and made available through specialist publications (some of which are used by ecotourists as guides to the bird and animal life), and even published on the Internet. Costa Rica is today perhaps the world's most advanced country in this respect. Legislation protecting its biodiversity maintains genetic wealth under its control, thereby generating businesses and income. Another interesting project carried out by INBIO is the contracting of local rural people, known as parataxonomists, for the collection of species samples in the park. Traditional local knowledge can thereby be used and preserved, at the same time that it gives local people an important alternative source of income.

Today, Costa Rica lives from its nature, with ecotourism now ranked second, after high-technology industry, as its chief source of income. In 1987, in recognition of the country's dedication to ecotourism and cooperation, Former President Oscar Arias was awarded the Nobel Peace Prize.

Nowadays, an enormous number of tourists from around the world visit the country's national parks, La Amistad among them, to delight in their natural beauty and see the country's immense biodiversity at close range. Costa Rica is showing us that the ethic of environmental conservation and cooperation can also be an international attraction, and hopefully, an example to the world.

La Amistad, together with other 40 protected areas in six countries, form the Mesoamerican Corridor, creating an ecological "bridge" joining the three Americas.

La Amistad International Park protects important headwaters in both Costa Rica and Panama, of fundamental importance for guaranteeing the water supply.

Flowers like the *Cirsum sp* fill the scenary at high montain. The park holds high levels of endemism.

In an area where political boundaries are losing their importance, capuchin monkeys (*Cebus capucinus*) delineate their territory with urine.

The color attracts, but at the same time signals danger. This small amphibian poison dart frog (*Dendrobates pumilio*) has a toxic poison, quickly able to paralyze the muscles of its prey.

The Resplendent Quetzal (*Pharomachnus mocinno*), the totemic bird of both the Mayan and Aztec cultures, represents freedom and wealth. At that time, killing a quetzal was a crime punished by death. Nowadays, unfortunately, it is rare to see the bird in its natural habitat.

Tropical ferns have existed here for 300 million years and are often used by traditional people for their healing properties.

An international park creates opportunities for many cooperative actions. This way, everybody wins: men and nature.

Nature shows off its perfection, the owl butterfly (*Caligo atreus*) simulates a huge eye on its wing to scare away predators.

In 1985, UNESCO declared the region a Transborder Natural World Heritage Site, encompassing the entire area of the international park, plus six surrounding protected areas, making a total of 570,000 hectares.

8 A history of evolution and revolution. Galapagos National Park.

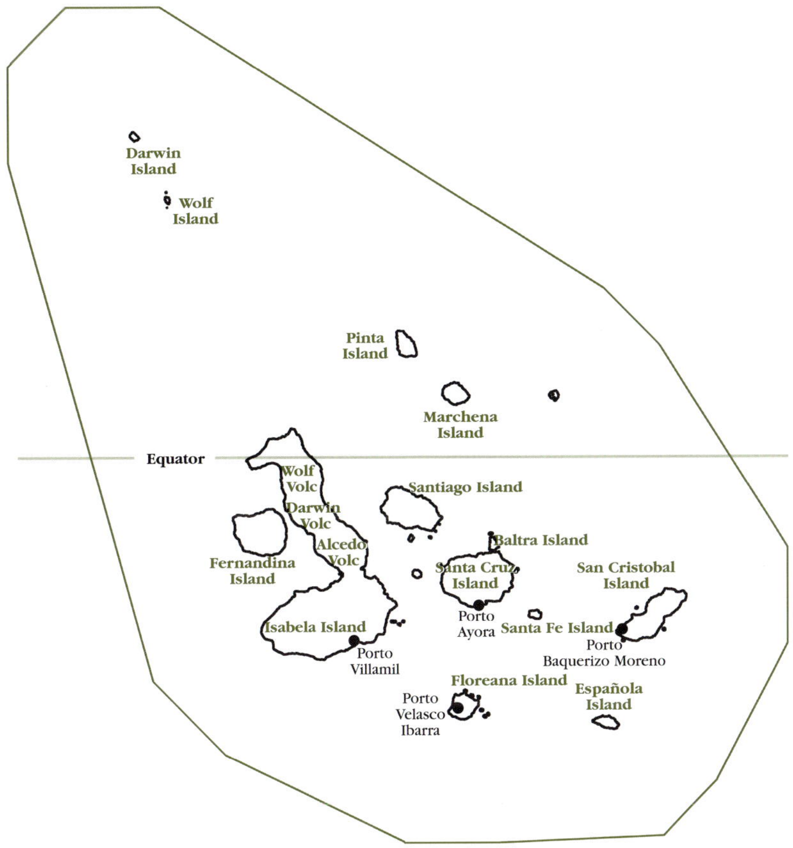

A 40-mile limit around the islands marks the boundary of the Galapagos Marine Reserve.

Location An archipelago in the Pacific Ocean, part of Ecuador's territory.
Area 800,000 hectares in the National Park and 13,300,000 hectares in the Marine Reserve.
Creation The National Park was created in 1959 and the Marine Reserve in 1998. The park was proclaimed a World Natural Heritage Site in 1978, extended in 2001 to include the Marine Reserve, and in 1985 was declared a Biosphere Reserve by UNESCO.
Ecosystem Arid plains in the low-lying land and cloud forest at higher altitudes on some islands.
Climate The islands have a tropical climate influenced by cold ocean currents, resulting in a dry, cool climate for most of the year.
Relevant features Tourists visit it to see the natural history of the Galapagos, monitored by guides who are also naturalists. The islands' communities take part in the search for solutions leading to sustainable management. Scientists carry out research aimed at conservation.

8 Galapagos National Park. A history of evolution and revolution.

The year was 1832. The English naturalist Charles Darwin was 22 at the time and had a weak stomach for ocean voyages. Upon boarding the HMS Beagle for an epic voyage of world discovery that would take five years and reveal the world to him, Darwin followed his father's advice and took with him a good supply of raisins. And that was all he was able to eat at the start. The crossing of the equator, however, brought the young naturalist renewed appetite. This strength probably came from the diversity of the forest that so fascinated him. As he himself would say in letters to his family and friends, begging pardon for his bouts of enthusiasm, "Describing the nature of South America to one who has never left Europe is like trying to describe colors to a blind person."

Darwin went to many of the places shown on this book, but one of them, in particular, would forever change his life, the history of science, and mankind's understanding of the origin of life. That place was the Galapagos Archipelago. There, besides seeing species never before imagined, he foresaw the very way in which life had evolved:
"I was so impressed with the organisms that I saw in the Galapagos (...) that I decided to blindly collect all kinds of data that could have anything to do with what species are. I read piles of books (...) Finally, some rays of light appeared. I am almost convinced that species are not (this is like confessing to a murder) immutable."

It was all there to be seen, life mutating in a constant process of adaptation to the environment. All that was needed was observant eyes and an open spirit, both of which Darwin possessed in abundance. The Galapagos consisted of 13 islands and over 100 rocks lost in the middle of the Pacific Ocean, almost 1,000 km from land, propitiating speciation. So much so that a large number of the reptiles and birds on the Galapagos are endemic, exclusive to the islands.

The islands had never been inhabited by Polynesians, or anybody else, except for rare visits from pirates, fishing boats, and whale hunters, beginning in the middle of the 17th century. They had therefore maintained very high indices of their original number of species (on some islands, still virtually untouched by humans, even today 95% of the original species remain).

During the time Darwin spent in those waters, he would record with wonder the land (its volcanic geology in constant eruption), the sea, which, due to the diversity of the currents, brought together animals from such different climates as penguins (*Spheniscus mendiculus*) and giant land turtles (*Geochelone elephantopus*). And, finally, the birds, mammals, and reptiles that, depending on which island they inhabited – wetter, drier, etc. – showed clear differences among them, as a result of the need to adapt to their specific environments, such as Darwin's finches, which received their name as a result of the naturalist's studies, and which became famous for exemplifying this process.

From one common ancestor, thirteen similar species of Darwin's finches have evolved on the islands, differing only in slight adaptations, mainly to their beaks.

The process continues today and has been studied for decades by Scientists Peter and Rosemary Grant, who became famous for documenting alterations in the size of Darwin finches' beaks, as new generations adapted to climatic or environmental changes. Evolution could be observed as it happened! The Galapagos were, as they would later be called, "a living laboratory of the evolution of species."

The Ecuadorian Archipelago is really a living laboratory, and within the reach of the eyes, or better still of tourists and their cameras. Having lived almost always far from human predators, the animals on the islands have no fear of humans. Today, the tourist visiting the Galapagos is able to enjoy a unique experience – that of observing wildlife close-up. Care has to be taken in walking the trails around the islands, so as not to disturb animals lazing in the sun or caring for their young, or gazing out to the sea as in the case of the sea lions (Zalophus californianus), the marine iguanas (Amblyrhynchus cristatus), and the many birds.

It is a veritable Eden. Currently, almost 80,000 tourists from around the world pay a high financial price to experience, even if only for a few days, their share of paradise.

Even with this immense natural treasure, the temptation to yield to mass tourism has never seduced the Galapagos. Ever since the national park was created in 1959, protecting 97% of the islands' territory, the impact of tourism has been carefully monitored. In 1974, this process was reinforced with the development of the Management Plan of the park, with the undertaking of numerous assessments of the environmental impact of tourism in each of the attractions of the park. This resulted in limiting the number of visitors on each tour, according to its capacity, so as to afford zero risk to the environment. For example, the routes of boats sailing among the islands are strictly controlled according to set timetables, affording maximum protection to marine life. In the same way, land trails were rerouted so as to create access to the attractions with minimum interference to the dynamics of animal life. The park grants credentials to bilingual naturalist guides, many of them native to the islands, who have a profound knowledge of the historical and environmental context of the region. Their job is also to make sure safety and conservation regulations are respected.

In the first years of operation of the park, the environmentalists and scientists concentrated on protecting just the terrestrial animals and ecosystems. However, with time and the invasion of the archipelago's waters by fishing boats from all over the world, measures had to be taken to protect the marine environment as well.

The decision led to controversy. After several attempts to implement a marine reserve, opposed by the local population who relied upon fishing for part of their livelihood, the Galapagos National Park Service (SPNG) and the Charles Darwin Foundation decided to reformulate the concept of the marine reserve. They therefore suggested a participatory process to include all the sectors involved in the preparation of a proposal. A "Nucleus Group" was set up with representatives from the fishing, tourism, and conservation sectors. Together, they drew up a Management Plan for the Galapagos Marine Reserve.

Dozens of meetings led to important decisions that served as a basis for the Special Law for the Galapagos, approved by the Ecuadorian government in 1998, creating the Galapagos Marine Reserve, which excluded industrial fishing within a 40 mile limit from the base line surrounding the islands (133,000 km^2 – the world's second largest marine reserve after Australia's Great Barrier Reef). The Galapagos were now doubly protected. Firstly, by new laws respecting the systemic linkage between the land and marine environments and, secondly, by a wide group of people, from local inhabitants to scientists, now directly involved in the conception, planning, and execution of the Management Plan. The Galapagos Special Law also regulated a series of questions that were fundamental to the environmental health of the islands, creating legislation to prevent and/or control, for example, the introduction of exotic species of flora and fauna. In the past, and continuing today, species of animals and plants brought in from outside have caused immense impact, due to their predatory action on native species. Nowadays, the islands have a special institution for vigilance known as SICGAL, with inspections both at exit points on the continent and at entry points on the islands. Human migration is also strictly controlled, whether of local inhabitants or of migrants. Economically, the Galapagos are better off than Ecuador itself, attracting a flow of migrants that seriously compromised environmental equilibrium. With the legislation of 1998, this question is now regulated, with the support of the majority of the local population. Nowadays, only registered residents, people who have married a local, or those who have working contracts may live on the islands.

Another example of synergy is the intense movement, not only of tourists, but also of donations, leveraged by the tourism sector. Hundreds of thousands of dollars are allocated to the park's needs and to the institutions that carry out the various programs directed to protecting the islands. The visitor tax of US$100 per head (with the exception of residents of countries of the Andean Pact and of the Mercosur trade blocs) generates a fundamental income of several million dollars per year. To get a better idea, almost half (45%) of this income remains in the park and covers almost its entire annual budget. A portion is also allocated to other institutions – among them the local government and the Navy – to finance environmental education programs and the protection of the Marine Reserve.

The positive results of the successful programs in the Galapagos also benefit the country as a whole. Each foreign tourist spends an average US$3,700 (2001 figures) on a package tour to the Galapagos, of which 50% go to continental Ecuador, 15% to the Galapagos, and 35% to the world economy. Almost half of Ecuador's total tourist revenue of a little under US$ 500 million derives from tourism in the archipelago.

The role of the Galapagos in attracting various actors in the conservation movement has been internationally recognized, inspiring other parks around the world to adopt similar strategies, integrating ecosystems, people, interests, and needs – in short, lives.

One of the fundamental institutions in this process has been the Charles Darwin Foundation, which was founded together with the park in 1959, and which, in addition to sharing strategic decision-making with the park authorities, acts as a bridge between the Galapagos and various international institutions, both in fundraising and in scientific research. Over the years, it has given logistical support to over 700 scientific missions, resulting in over 600 research papers being published on the archipelago and/or on its species. The Galapagos Islands are today one of the best known and most studied places on Earth, although much still remains to be done. This knowledge has always been an important source of information for planning the actions and policies of the park itself.

Such productivity of ideas, initiatives, and of nature itself, will nourish the legend of the Galapagos for a long time, fascinating generation upon generation after Charles Darwin. Perhaps the mystery that so enchants visitors comes from the volcanic intensity of the islands and from the unforgettable images of lava pouring into a boiling sea. Perhaps it comes from the peculiarities of some of its unique animals: the giant turtle or the marine-iguana, animals that seem to have come directly from a Jurassic past. Or maybe it is the fact that these islands inspired one of the most important theories of the modern age, the one of the evolution of the species. Perhaps it is all these together, given that a legend is made of many facets, and in the Galapagos, each expression of the legend is absolutely unforgettable.

The Galapagos National Park is a natural history lesson "in loco". The abundance of wildlife is such that the laws of nature can easily be observed. The brown pelicans (*Pelecanus occidentalis*), for example, stay close to the sea lions (*Zalophus californianus*), waiting for a chance to grab some leftover fish.

Hundreds of sea lions rest on the beaches, oblivious to the presence of humans. The beach belongs to them, and we are the intruders.

The swallow-tailed gull (*Creagrus furcatus*), contrary to other birds in the region, feeds from the ocean at night. During the daytime, they remain in their nests on the rock faces.

With the gradual increase in predatory fishing of the surrounding seas, environmentalists and scientists turned their attention to protecting the marine ecology of the region, creating the Galapagos Marine Reserve.

Cactus forests dot the arid, volcanic landscape of the Galapagos.

On the island of Española, an illustrious visitor rests. The rare waved albatross (*Diomedea irrorata*), the largest bird of the region, can be seen in its nest. The region is the only place in the world where it nests.

The volcanoes on the islands of Isabela and Fernandina are among the world's most active. Explosions of lava make the sea boil and constantly reshape the islands.

The male magnificent frigatebird (*Fregata magnificens*) inflates its red-colored chest and performs elaborate dances to attract the female. The ritual can require two weeks of dedication until the partner finally accepts.

Ecuador's first national park, the Galapagos Archipelago fights for its preservation – an arduous and intense task when a fragile, isolated, and unique ecosystem comes up against the pace of modern life.

Able to remain underwater for over an hour, marine iguanas (*Amblyrhynchus cristatus*), endemic to the region, rarely dive for more than ten minutes, due to the cold waters of the region.

Despite its delicate appearance, the red-footed booby (*Sula sula*) dives into the sea at over 100 km per hour and up to 10 m deep in search of a catch.

Due to the constant care taken by well-trained park rangers and tourist guides, the animals of the Galapagos, like this brown pelican, are not disturbed by the presence of humans.

The future for national parks

Cecília Ferraz

The National Parks of Latin America should serve, in the near future, as a world reference, not only in terms of scenic beauty, but also in terms of their activities in conservation and management of biodiversity and in the implementation of sustainable activities integrated with the surrounding population. Everything points towards this route being taken.

However, it should be remembered that this will only happen with the help of pressure from society as a whole. Society has the right to benefit from these areas. A desire to seek nature is becoming ever more preeminent.

For this reason, I believe that the national parks of Latin America will be increasingly the target of pressure for good governance from, firstly, tourists and then from society as a whole in terms of how they are implemented. This implementation deals not only with actions necessary for research and management of the areas, but also with greater and improved services in welcoming and supporting visitors, and also with activities with the local inhabitants in their respective surrounding areas.

Several of the cases presented in this book indicate the probable paths to be followed and experiences to be replicated. The publicizing of activities that are now being practiced in several parks is essential at this time.

For example, in my opinion, the first and foremost condition for successful implementation of a national park is the staff, which should be composed of both technical professionals and of park rangers with specialist training in the management of national parks. After this condition is achieved, we should also have an effective and efficient system of planning, regularization of land ownership, partnerships, participation, and, of course, financial resources.

Cecília Ferraz is Director of Ecosystems at IBAMA.

National parks and other protected areas

Cláudio C. Maretti

Protected areas are, at one and the same time, both an example of success and an enormous challenge.

The challenge must be faced, as the protected areas are one of the best instruments for reaching the objectives of nature conservation, as an integral part of the processes of sustainable development. The degree of success can be seen in the figures of the 2003 edition of the United Nations List of Protected Areas – these already cover 11.5% of the Earth's surface. Currently, however, more than fix parameters for their creation, we need to protect important samples of biological diversity and culturally significant natural attributes. This is where the challenge begins. Although a lot of progress has been made, there are types of environment that are not well-protected, particularly aquatic ecosystems. Only around 0.5% of the seas and oceans are protected worldwide. Neither is protection adequately distributed around the globe. In Brazil, for example, lack of protection is evident in the Caatinga, in the Araucaria Pine Forest, and in the Cerrado.

Ever since areas were first reserved for specific purposes, many centuries ago, the objectives have varied, as have management methods. At a certain time, national parks were held as models, influencing many of the conservation units we have today. Initially, the creation of national parks aimed at protecting areas of outstanding natural beauty and at promoting touristic purposes. However, the origin of national parks is also linked to the relationship betwen countries and their territories. That is, they were national symbols aimed at demonstrating a nation's effective control over its territory.

Nowadays, in Brazil, parks can be national, state, or municipal. Together with other categories of conservation units (such as ecological stations, extractive reserves, environmental protection areas, etc.), they should contribute, primarily and by definition, to the conservation of nature. In the case of the Amazon, for example, its national parks and other categories of conservation units, act as fundamental elements in preventing deforestation. Nevertheless, even today, parks still serve the purpose of tourism as well. In terms of international classification, what defines the type of management (or usage) is its set of objectives. It is, therefore, only through tourism that parks can totally fulfill their functions. And this is where one of the most important aspects of this category is found: in addition to respecting the rights of people to visit natural areas, it is understood that they will also increasingly come to understand and defend nature. And this is reinforced by the mythic value that an area acquires after being declared a national park and becoming a beacon for tourism – for better or for worse.

Nowadays, however, there is a new paradigm for the management of protected areas – established at the World Parks Congress (2003). More than just being areas of 'reserved' land, protected areas are expected to fulfill the functions of sustained development, as well as help to 'give shape' to the development model and type of scenery that society wants. They are no longer conceived and created in isolation, but rather as part of systems, networks, and mosaics, in addition to being integrated with their surrounding regions. What's more, it is no longer just the conservation units created and administered by national governments that count, nor are they viable if simply imposed technocratically. On the contrary, they are created de facto by a variety of partnerships, aiming at integrating, through participatory mechanisms, a multiplicity of capabilities needed for an adaptive management.

In this context, national parks must be stripped of their mythical status as isolated and overvalued beings, able to resolve the question of protecting nature alone.

They need to be understood as valuable in themselves, however, they can only reach their full worth once inserted in systems of protected areas and integrated within their regions.

It has already been recognized that, more than just defining a space, or protecting an asset or a fundamental natural quality, a national park represents the formal definition of the value (necessarily cultural) that a society attributes to this area and its natural resources – as well as to the cultural values associated with them. It is, therefore, important to understand them as product of our appreciation of the nature. It is our posture that is important: the protection of nature must be integrated with all our actions –including plans, works, industries, urbanization etc. Just as it is necessary to consider protected areas as fundamental elements in the ordering of our territory, in the improvement of our landscape, and in the organization of our development. However, in order to well implement national parks, and other categories of protected areas, it is essencial to have the necessary resources – the financial ones, among others. And it is also essential to have sufficient and well-trained professionals, in addition to the support of social players, especially local ones, for a park to survive in its functions.

Cláudio C. Maretti collaborated on this book with suggestions and comments, especially technical ones. He is a doctor of human geography and regional vice president of the World Committee for Protected Areas of the UICN and coordinator of the Protected Areas Program in addition to giving support to the Protected Areas Program of the Amazon (ARPA), at WWF-Brazil.

Interview with Luciana Napchan

The book is a five-year project. What does it feel like to have completed it?

This book has been an exercise in patience and perseverance. During the process, I was fortunate to be able to experience the richness involved in its production. Just as taking each photograph was a small metaphor for life, so was the making of this book. A lot of perseverance was needed to persuade all those involved to believe in such a long project.
For me, it was also an opportunity. Photography allows me to communicate, silently, the way in which I perceive nature. Nevertheless, I wish to make it clear that we do not need to go to places as far away as those in the book to see the wonders of the natural world. All we need is to look at the trees in our own street, observe the perfect equilibrium of a small plant in a vase, admire the swift passage of clouds on a windy day, or be enchanted by the feline grace of a neighborhood cat.
Besides the opportunity to communicate, the book is also an opportunity to scream. Due to my passion for nature, I feel the need to cry out trough my photography, as the trees and animals are unable to do it for their own conservation. They just die and disappear.

Therefore, does photography have many meanings for you?

When I look at a photograph, I always think of the reality that the photographer was unable to include. What was happening outside the frame when the photograph was being taken? Were there people around, or behind the camera? Did they all leave afterwards? What was happening just before the photo was snapped? What was the weather like at the location at the time of the photograph? What time of the day was it taken? How long did the photographer take to get there?
A photograph immortalizes 1/250 of a second of reality. Many decisions and choices have to be made before every photograph.
So, I would summarize each photo as being a present given to the photographer by his own insistence.
The nature photographer has to be stubborn. The final shot is a positive conjunction of various factors that do not occur by chance. One must be at the right time at the right moment, with the right intention. In order to do that, one must study the situation to be experienced. Pinches of sensibility, intuition, adaptability, and courage are indispensable. I had the chance to experience all this while taking the images for this book. By the way, this book is the result of a desire to give something in return, to give back to nature everything that it has given me, and therefore, help to raise awareness of preservation.

Your work has always been classified as "artistic" photography or interpretations of reality. In this book, showing, as it does, photographs of nature, the main objective is fidelity to the nature portrayed – precision. In terms of the way you see it, what has changed?

Nature is so perfect. It has no need of interference. At the moment that I am photographing it, it's like a relationship of devotion. I prostrate myself before it. Nature is my teacher and I am there to learn. Therefore, every time I enter a forest, I always ask for permission silently. National parks and other protected areas, the last preserved sites of the planet, are sacred places. Each millimeter of a century-old forest is sacred. By photographing this sacred nature, my aim is to bring it closer to people, so that we can see how it is part of us, and we are part of it.

How do you see photography and the question of environmental protection? Can it be a tool for transforming reality?

Taking photographs, which is a continuous act of looking, selecting, and focusing, teaches us to see with greater precision. By constantly taking photographs, we become more sensitive to the subtleties of light, shades, and details. Therefore, I believe that photography can be an act of environmental perception.

In today's world, we should be both economical and ecological. Non-recyclable waste can be found everywhere on a daily basis, including the chemical and/or technological waste resulting from photography. Consequently, by linking our art and profession with something in which we believe, it becomes more powerful, positive, economical, and ecological.

Photography brings to people that which is distant or which they have never experienced. It can be a great tool when used in favor of environmental protection.

Which of the places you visited most affected you, and why?

Making this book entailed discovering Latin America. In Brazil, we have always faced the ocean with our eyes turned toward Europe and North America. We idealize everything that comes from them. And we remain blissfully unaware of Latin America.

What was fascinating producing this book was discovering the similarities among the people of Latin America, with their kindness and hospitality.

By getting to know these places, I strengthened the idea that nature has no borders. Similar species to those of Brazil's Atlantic rain forest, for example, can be seen in the forests of Costa Rica. I had the chance again to experience the unity underlying everything.

How did you see and portray the human question – the human being – within these protected areas?

We are all part of nature. There is nothing on this planet or in this universe that is not part of nature. Nevertheless, we do not always live in harmony with the nature that surrounds us.

However, in this case the word harmony fits like a glove. On my travels around the national parks, I saw men, women, and children in harmony with the environment in which they lived. They perceived the divine in each piece of land, in the creeks, in the birds, and in other human beings.

You need humility in order to respect nature, as you must equally respect the ant, the jaguar, the mosses, and

the magnificent trees. The lesson learnt on my travels was one of humility. That is what I tried to capture, together with the dignity that I saw in the eyes of many of those I photographed.
Another significant detail was that most of the people with whom I walked during this journey, or whom I photographed in these protected areas, were there because this was the way of life to which they aspired and which they loved. The park rangers in "Torres del Paine", for example, live isolated lives and are passionate about their mountains. They see themselves as, and indeed they are, the guardians of these wonders, making sure their harmony is not disturbed. They really are attached to the environment in which they live.
And, as opposed to material goods, it is a rewarding experience to become attached to nature. Attachment generates love, and by loving a landscape, an animal, a river, or a tree, we create an involvement that in turn generates a possibility of preservation.

Do you feel that photography has the power to transport us to these other realities, co-opting people through observation?

I believe that images have the power to make us forget our own reality. Images can make us dream or weep. Images allow us to travel. There is a very beautiful text by Roland Barthes that defines a good photographic image, "To really see a photo, better we raise our heads and close our eyes – in silence, eyes closed, letting each detail rise alone to our affective consciousness."
A fine image touches us through affection, through emotion.
The photographer can choose between enlarging an aberration or perfection. Environmental awareness and preservation must focus on both destruction and conservation. Both work as forces for mobilization.

Photographing animals in nature is not easy. How did you do it?

In some situations, establishing a relationship between the photographer and the subject can be useful in photographing animals. It is almost like a love affair. In the Manu National Park, in the Peruvian Amazon, for example, I spent hours following a howler monkey until he slowly became accustomed to my presence, looking at me and at the camera, until one day he climbed onto my shoulder, and we became friends. It is fascinating, as in doing so, we get the chance to learn a little of the animal's personality.
Each animal I photograph becomes unique to me. In general, they are curious and depending on the species, they will often realize I am there, close by, and well intentioned. Like us, animals have their daily routines. The photographer or observer who wants to see them should be patient and wait.
In one of my trips, I spent days in a small hide, made out of palm fronds, expecting the right moment to take a picture of the fauna. The place was where spider monkeys came to feed. There was room for

a small stool, and there I waited in silence. After a few days, I decided to take a notebook and a book. I read an 800-page book over some days. The monkeys did not appear, though. Frequently, images are dreamed of but never taken.
I remember an occasion when I was walking in the Amazon Forest with an indigenous guide. I carried my photographic equipment and he, who usually brought his bow and arrow, did not have his weapons that morning. We began to hear a deafening noise. It sounded to me like a pack of wild dogs, and following the noise came a strong smell. It was, in fact, hundreds of wild pigs. My guide shouted out, "Wild pigs, run away." I hesitated, imagining the great photo this could make …but I thought, "He is a native and wants to flee!!!! If he is afraid, then this has got to be serious…." So we fled and he yelled, "Climb a tree" and I couldn't even imagine how… I had forgotten to learn how to climb trees. Days later, I was telling the story at the lodge where we were staying, and the park rangers and guides said, "Where there are wild pigs, there are jaguars hunting them, and behind the jaguar you can see a rare bird that feeds on the organisms living in the jaguar's fur. You missed an extremely rare opportunity." I saw, by the look in their eyes, that a new awareness had come about with the creation of the national parks, where nature is no longer seen as the enemy of mankind.

What was it like to be a woman and alone in the midst of the bush for so long?

National parks are, in theory, places that are reasonably well protected, although one must be aware that as a woman one needs to be always alert to situations that could become uncomfortable. Anyway, regardless of gender, one should always be gentle when entering a protected area.
For a female photographer, in some situations, doors open with greater ease. A woman entering a small village alone is less threatening than a bearded man. As a woman, I am able to frequent most of the women's quarters and am also received courteously by the local men. This way, I am able to achieve a closer intimacy with the subject of my photograph.
But, I fear mankind more than I fear animals! There are moments when I need to remember that life is more important than art.

Luciana Napchan

Luciana Napchan studied film at the University of São Paulo, Brazil, and at the Munich School of Cinema, Germany, where she worked as an assistant for European photographers. Upon her return to Brazil, she worked as a photojournalist at the Folha de S. Paulo newspaper and for various magazines, such as Marie Claire, Caminhos da Terra, and Vogue.
Her photography focuses mainly on themes related to social and environmental issues. She has held exhibitions at museums and cultural centers in Brazil and abroad, combining photography and fine arts.
Her work "Gritos da Rua" (I Scream) was exhibited at the Art Museum of the Americas in Washington DC, USA, and published in Time Magazine in a special issue on Young Leaders for the New Millennium.

Interview with Walter Behr

Which sector of the public are you aiming to reach with this book, and how can it contribute to involving more people in the environmental movement?

Global warming and the increase in natural disasters are just some of the environmental reactions to our unsustainable development, which already consumes 20% more of our natural resources than our planet can replace. These environmental reactions already influence, and will increasingly continue to influence, the daily lives of everyone who, whether they want to or not, will have to become more seriously involved in the matter, even if only for the sake of survival. I always ask myself how long does humanity plan on living on this planet with a reasonable quality of life? For another one hundred or one thousand years? Judging by how voraciously we consume our natural resources, I would say that we seem to have already decided on shortening our stay here…

However, decisions rest with a small elite, to the detriment of billions of people and of future generations. Are we not going to ask the children what they want? Aren't we going to ask everyone who is involved? The influence of those people who hold economic power – in general concentrated in large urban centers – is vital in this process. Saving the Amazon, for example, depends equally, or more, on people in São Paulo and Brasília, where economic power and decision-making is concentrated, than on those people in Manaus or Belém. It is the awareness of people living in our larger cities that this book intends to raise. In Brazil, as in Latin America in general, people are not familiar with what constitutes a national park and the relevant functions it performs within an environmental and economic context. The publication of this book is intended to bring activities carried out in our national parks to the notice of the public. That is why we have chosen examples of successful environmental programs which, in addition to inspiring the administrations of public and private protected areas, can awaken the interest of those who are not yet part of the community of people enjoying the wonders that a national park can provide.

How do people see national parks?

A national park maintains a natural environment and aims to provide environmental education, protection for its biodiversity, scientific research, and public visitation.

But practice is still a long way from theory, unfortunately. With rare exceptions, national parks are seen as "no man's land", where anyone may do what he/she wants. Or, often, people visit a national park, but do not realize that the area visited is a park, as it happens with 90% of the 750,000 people who each year visit the Sugar Loaf, located within the Tijuca National Park.

Society as a whole has to feel responsibility toward these areas and participate directly in the development of programs that enable the achievement of the objectives of a park.

By leaving them only in the hands of the government, both protected and "unprotected" natural areas will continue to be

"no man's land", or ignored, as in the case of the Amazon, which is being rapidly destroyed.
We need to act, each person doing his or her part, before it is too late. So, the book aims to show a few examples that can be replicated, and not just in protected areas.

You mentioned the Amazon. What can be done to avoid total destruction of the area?

I had an interesting experience in the Amazon. In 2003, when I became part of the Executive Management of Tefé (IBAMA), whose area of operation covers over 50 million hectares of well-preserved forest in western Amazon, there were another 14 new environmental analysts working for IBAMA. Working conditions were not conducive to doing a good job. We were, however, a small enthusiastic group of professionals, among which, forestry engineers, biologists, administrators, an architect, and even a specialist in ecotourism who had graduated from the London School of Economics. So, we developed an emergency plan, costing under US$ 600,000, that foresaw providing the management with the minimum infrastructure to make a start on changing the situation of neglect in this immense region. We took our project to the heads of IBAMA in Brasília and argued that if this kind of strategic support was not offered, within a year all the environmental analysts would end up leaving, since staying in an isolated area without even the minimum working conditions was unthinkable. Nothing was done, and, in fact, without the necessary infrastructure, everyone did leave within a year, and a great opportunity was lost.

A friend of mine once made a comparison, saying that Tefé ought to be a sort of NASA of the environment, as it harbors the most preserved part of the Amazon and is of strategic and planetary importance. For less than US$ 600,000 we lost this great opportunity.
You asked me what could be done. This is a small example of what can be done. Before long, that same administration will be receiving a new group of environmental analysts coming from another rare public service exam.
Will we lose this great opportunity again?

How did you become involved in the environmental movement?

My parents brought up my brothers and me in close contact with nature. This was, without a doubt, the determining factor of our lives. After graduation, I was faced with a dilemma: whether to dedicate the most important part of my life to earning money and leading a conventional life, or whether to fight for the environmental cause.
My friends tried to advise me to first guarantee my economic stability before engaging in the environmental struggle. As I might well have met my death before ever reaching that day, I decided to get involved straight away.
I went to live in the Serra da Bocaina mountain range, and helped set up an environmental NGO. I am now an environmental analyst in the Serra da Bocaina National Park.

Does one have to give up everything and radically change lifestyle in order to become part of the environmental movement?

No. Small changes in attitude can help a lot, as can assessing the consequences of our decisions. The bottom line is that we must be critical. For example, should we buy timber from the Amazon or from a reforestation when we want to build a house? Should we eat less meat and more organic products, or produce less waste? I do not want to be extreme, but each and every one of us should be aware of the consequences of our actions in various aspects.
By consuming meat, for example, are we aware that 70% of the deforestation in the Amazon is in order to create cattle pasture? This kind of questioning brings about a new relationship with our planet. We begin to feel a responsibility toward it and to change our attitudes.
Another question that people tend to avoid is what the responsibility of each one of us for the social inequality in which we live is. And we are creating ever more huge inequalities. There is an extreme degree of selfishness in thinking that others do not have the same rights as we do to enjoy a similar degree of well-being, as it is simply environmentally impossible for all to have the same standard of living as the elite's. Environmental degradation is the direct result of human misery: one leads to the other. And when one becomes involved in the environmental struggle, you begin to see this relationship and to try to build a better, more equal world.

But the environment, on the other hand, does not lead us to action, but rather to the contemplation of the beautiful...

Without a doubt, it is important to contemplate the beauty of this earthly paradise that has been offered to us and stop putting so much energy into its destruction. Obviously, we need to go beyond mere contemplation and act. We must go beyond the ease of contemplation and have the courage to put thoughts into action, as our book suggests. We are so daring in our acts of destruction. There is no lack of funding and subsidized interest rates for expanding soy plantations, right? We, on the other hand, should all dare to protect and care for nature, only using it when necessary, and with great care.
Or else we will keep on subsidizing soybeans and cattle, for example, with continued environmental destruction and social misery. We, in Brazil, possess one of the few expanding agricultural frontiers in the world. After destroying the Cerrado, we are advancing ceaselessly into the world's last great remaining natural area, the Amazon. The media, in general, broadcast misleading information. Consider once more the question of soybean plantations and cattle ranching. We exult over the trade surplus afforded by our agricultural exports. Is Latin America likely to be saved by selling agricultural commodities subsidized by the destruction of our natural resources – our soil, water and tropical forests?

So, how did the idea for the book come about?

Ever since the creation of the Pro-Bocaina Association, I began to get involved in the Serra da Bocaina National Park, and always faced great difficulties – not to say obstacles – in putting actions aimed at changing its state of neglect and destruction into action. Then we decided to investigate successful examples in Brazil and in the rest of Latin America that could serve as inspiration. The wealth that we found came as an unexpected surprise, as even today, it is usually examples from North America, coming from a completely different cultural, economic, and environmental background, that are held up as models for conservation at seminars and congresses.

After seeing examples in Latin America that are imaginative and successful, and in some cases examples to be followed by the rest of the world, we were motivated to put them into a book. We principally chose examples that had the greatest chances of replication here in Brazil. That is how the idea for the book arose.

Did Latin American national parks impress you?

Brazilians hold neighbors in low regard. Latin America is a continent blessed with incredible natural and cultural beauty. Even today, in the 21st century, there are indigenous people who have not had any contact with modern civilisation. We have a history of great cultures such as the Inca's. And we also have imaginative examples of environmental management.

I was impressed by the examples I saw of national parks in Costa Rica, Chile, Argentina, and Ecuador, among others portrayed in the book. We have a lot to learn from each other, exchanging information, and, more than ever, moving into action and getting down to business. After doing the book, I have concluded that this is what we should do. We have had enough talk. It is now time to act.

But environmental wisdom, until recently, preached that national parks should be left in isolation and untouched.

Yes. There was resistance to the participation of society in national parks and fortunately this has changed significantly. Now we need to facilitate and encourage partnerships with NGOs, private sectors, research institutions, volunteers, trainees... In Brazil, the national parks should form their own consultative councils, another important tool in which the public can participate. Local governments, NGOs, and people living in areas surrounding the national parks need to be involved. Many landowners in the surrounding areas, with the right incentive, end up taking part and even creating their own private protected areas, known in Brazil as the Natural Heritage Private Reserves (RPPNs), wherever the law permits. Expanding protection to surrounding areas is strategic to the success of conservation units.

And what needs to be done next regarding the Brazilian national parks?

In my view, there are various factors. I have already mentioned the importance of partnerships with NGOs, private sectors, and universities. Pressure on governments by society is also important. The more the public in general becomes involved, the more attention is given to protected areas. Another factor omnipresent in Latin America is the lack of funding. I recall visiting a national park in Germany, where they had a budget of US$ 4 million available just to build another visitor center... The park even had its own printing house for its publicity material. This is very different from our reality in Latin America.

In Brazil, however, with the new legislation introduced in 2000 instituting the National System of Conservation Units (SNUC), at least 0.5% of the value of enterprises that cause environmental impact must be allocated as environmental compensation to protected areas.
These funds could change the history of neglect in Brazil's conservation units.
Absolute priority should, therefore, be given to using these funds, right? Unfortunately, this is not happening. We are wasting time with this issue, to the point that powerful lobbies have already been formed to contest the values of these environmental compensations, and attempts are being made to have the law changed so that companies pay less.

What are your comments on the direct use of protected areas?

There is a Brazilian model that I hold in high regard, the extractive reserves. There are few examples in the world like that. The reserves resulted from the efforts of rubber tappers, and the ideas and personal sacrifice of Chico Mendes. This category of protected area guarantees the continued presence of traditional people within the unit and promotes the sustainable use of its resources. Its creation and planning are made by the local community, given that representatives of the local inhabitants form the majority in the deliberative councils that make the decisions. This accounts for their great potential for success. It is civil society taking responsibility for the process, and it has been a great Brazilian example of how to limit the destruction of nature by involving local population in its protection.
I had an experience of over one year in the Extractive Reserve of the Lower Juruá (Reserva Extrativista do Baixo Juruá), in western Amazon. Less than 10% of its area is used for resource extraction or production. From 80% to 90% of its area is preserved, even taking into account the border effect. When its resources are threatened by incursions, the inhabitants of the reserve defend their territory or seek help from the public authorities. In Juruá, it was the community itself that built the IBAMA headquarters and welcomed its officials. The reflexes of this successful experiment have ended up spreading to adjacent areas and providing the incentive for other communities to request the creation of further extractive reserves.
In terms of national parks, I feel we have much to learn from our neighbors, judging from my experience. In terms of direct use of reserves, however, the examples of Brazil, of the Amazon, and of the riverside dwellers have a lot to teach.

Walter Behr

Walter Behr graduated in Business Administration from the Getúlio Vargas Foundation School (FGV-SP). At the beginning of the 1990s, he worked for Greenpeace and in 1993 was one of the founders of the Pro-Bocaina Association, taking part in the preparation of the Management Plan for the Serra da Bocaina National Park.
In 2002, he joined the Brazilian Environmental Agency (IBAMA) as an environmental analyst. Until 2004, he worked in the Lower Juruá Extractive Reserve in the Brazilian Amazon. He is currently responsible for environmental compensations and for partnerships at the Serra da Bocaina National Park.

Bibliography

ADAMS, Ansel. *Examples, the making of 40 photographs*. Little, Brown, 1983.

ALTMAN, Nathaniel. *Sacred trees*. Sierra Club Book, 1994.

BARBIERI, Gian Paolo. *Madagascar*. The Harvill Press, 1995.

BLOSSFELDT, Karl. *Photographs*. Taschen, 1994.

CALLIGARIS, Gustavo. *Patagonia, el país de las visiones*. Patagonia Geográfica, 2002.

Cerrado, Vastos Espaços. Edições Alumbramento, 1992/1993.

CHAPE, S. et al. (compilers). *2003 United Nations list of protected areas*. Gland e Cambridge: IUCN e UNEP-WCMC, 2003.

COUTINHO, F. & Brito et al. From Pre-assumptions to A 'Just World Conserving Nature': The Role of Category VI in Protecting Landscapes; in: Brown, J. at al *The protected landscape approach: linking nature, culture and community*, 2005.

CURTIS, Edwards S. *Native Nations*. Callaway Editions, 1993.

DE ROY, Tuy. *Galapagos: Islands born of fire*. David Bateman, 1998.

DPNG. *Plan de manejo de conservación y uso sustentable para la Reserva Marina de Galápagos*. Dirección Parque Nacional Galápagos. Quito; Registro Oficial nº 173. 1999.

FERREIRA, L. V. et al. *Protected areas or paper parks: the importance of protected areas in reducing deforestation in Rondônia, Brazil*. WWF-Brasil e Inpa-PDBFF. 2001.

FUNDACIÓN NATURA. *Informe Galápagos*, 1999-2000 e 2000-2001.

GARAY N, Gladys et al. *Conociendo la flora y montaña de Torres del Paine*. 1997.

GOELITZ, J. *Secrets from the lives of tree*. Planetary Publicatons, 1991.

GRANT, Peter et al. *The beak of the finch*. J. Weiner, 1994.

KEMF, Elizabeth. *The law of the mother, protecting indigenous peoples in protected areas*. Sierra Club Book, 1993.

KRISHNAMURTI. *Sobre a natureza e o meio ambiente*. 1991.

LANTING, Frans. *Jungles*. Taschen. 2000.

MACQUARRIE, Kim. *El paraíso amazónico del Perú: Manu – Parque Nacional y Reserva de la Biosfera* (edição bilíngüe). 1992.

MARETTI, C. C. Conservação e valores: relações entre áreas protegidas e indígenas, possíveis conflitos e soluções. In: *ISA*. *Desafios Socioambientais: Sobreposições entre Terras Indígenas e Unidades de Conservação no Brasil*. São Paulo; Instituto Socioambiental, 2004.

OYARZÚN, Gaston et al. *Torres del Paine*. Editor Dominique Verhasselt, 1997.

PHILLIPS, Adrian. Turning ideas on their head; the new paradigm for protected areas; in: Jaireth, Hanna & Smyth, Dermot (eds.) *Innovative governance: indigenous peoples, local communities and protected areas*. Nova Délhi; Ane Books, 2003.

PORTER, Eliot. *Galapagos: the flow of wildness*. vol. 1, Discovery – Sierra Club, Ballantine, 1968.

SÁ CORRÊA, Marcos et al. *Parque Nacional da Tijuca. 140 Anos da Reconstrução de uma Floresta*. Ouro sobre Azul Editora Ltda., 2001.

TERBORGH, John et al. *Making parks work*. Island Press, 2002.

WILSON, Edwards O. *O futuro da vida*, 2002.

Acknowledgements

Many people and institutions have helped in the production of this book, whether with information, goodwill or on-site support. Since this is a project that has lasted for over five years, we hope nobody has been forgotten – to all of you, our thanks.

Special acknowledgements to Guilherme Leal – who believed in this project right from the start –,Luiz Seabra, Pedro Passos, José Eduardo Dutra, Wilson Santarosa, Izeusse Braga, Luiz Carlos Burti and Leandro Burti – who made this book possible.

Our special thanks to: Antonio Portinho, Camila Alves, Catherine Kuglik, Cecília Ferraz, Claudia Guimarães, Claudio Edinger, Claudio Maretti, Daniel Di Giorgi Toffoli, Deborah Neistein, Dorothy Napchan, Efraim Zaclis, Fábio Cesnik, Fábio Feldman, Fernando Cavalheiro, Fernando Furriela, Fernando Zuffo e Adriano, Flavia Tanaka e Benê, Flavio Ignacio, Geraldo Falcão, Guilherme Farah, Heloisa Vasconcellos, Henrique dos Santos Pereira, Hugo Penteado, Ilan Wallach, Ioram Finguerman, Jairo da Rocha, João Araújo, Justo Alcazar, J. C. Pierina, Leonardo Muniz, Luciana Ferreira Leite Pinto, Marcelo Sanovikcz, Marcelo Soderi, Marcos Egidio, Maria Berenhoc, Maria Luiza, Maria Paula Fonseca, Mauro Berenhoc, Nilton Sonoki, Paulo Belleza, Paulo Sergio B. G. do Nascimento, Raquel Biderman, Renata Paiolli, Renata Sbardelini, Rita Vilarim, Rozely Ferreira, Rudi Böhm, Samuel Napchan, Silvio Pereira, Socorro, Stefania Valle, Stela Grisotti, Tabata, Toninho, Vanessa Kaufmann and our respective families.

Manu National Park:
Inrena – National Institute for Natural Resources
Manu National Park Administration
Peru Verde
Inka Natura Travel
Association for the Conservation of the Southern Forests (ACSS)
Pantiacolla Tours- Marianne van Vlaardingen
Special thanks to Charles Munn, Mariana Valqui, Modesto, Walter Mancilla, John Terborgh, Daniel Blanco, Rafael Tamasteiro, Juan Carlos Flores, Jesus, Raul, Rùben, Carlos, Patricia and Regent Gravel

Torres del Paine National Park
Conaf – National Forest Corporation
Torres del Paine National Park Administration, Guillermo Santana
Big Foot, Sergio Echeverria
path@gone, don Alfonso
Hosteria Lago Grey
Travesia Pehoe, Johny Maclean
Special thanks to Don Oscar Guineo, Patricia Perez, Park Rangers Freddy Bariento, Jorge, Juan Jose Salaz, Carlos, Alejandro and Ulisses

Chapada dos Veadeiros National Park:
IBAMA – Brazilian Institute for the Environment and Renewable Natural Resources
Chapada dos Veadeiros National Park Administration, Rosa Lia Gondim, Pedro Alberto Bignelli, Vilmone Manoel Ferreira
WWF – Worldwide Fund for Nature/ Veadeiros Project, Ricardo Mesquita da Fonseca, Roseli Madeira, Luiz José do Rego da Cunha Lima
OCA – Office for Science and Art, Paulo Maluhy
Chapada dos Veadeiros Visitor Guide Association, Valdoram Bastos
Special thanks to Lula and Yashen

Arenal Volcano National Park and La Amistad International Park (Costa Rica).
MINAE – Ministry for Energy and the Environment
SINAC – National Conservation Areas System, Francisco J. González, Vera Violeta Montero, Francisco
Amistad Conservation Area – Pacifico, Gladys De Marco
La Amistad National Park Administration, Ronald Chan Fonseca
Arenal Volcano National Park Administration, Gilberto Chávez Matamoros
Inbio (National Institute for Biodiversity), Angel Solís, Karla Sanabria M., Vilma Obando Acuña, Mario
OTS – Organization for Tropical Studies, Rodolfo Quirós, Claude J. Tremblay
Monteverde Conservation Association, Ronald Chavarria Viales
Children's Eternal Rain Forest
Ohoje Conservation Organization
Tabacón Hot Springs Resort & Spa
Sky walk, Monteverde
Danaus Ecocenter
Special thanks to Lola and Maurício Frajman, Helen and José Miremberg, Cyra Mikowski, Hugo Solano and Johana Padilla

Nahuel Huapi National Park:
National Parks Administration, Salvador Vellido
Nahuel Huapi National Park Administration: Sergio Rusak, Victor Arrechea, Alberto E. Quintero, Claudia de La Orden, Claudio Chehébar, Maria Soledad Caracotche, Marcia Urbinatti.
Park Rangers Association of Argentina, Daniel Paz
Vida Silvestre Foundation, Alejandro Vila
Fire Department, Forest Fires
Llao Llao Hotel & Resort Golf Spa
Hosteria Las Balsas, Villa La Angostura
Panamericano Hotel
Hotel Tronador
Hotel Design Suites & Towers
Cabañas Rosas Amarillas
Puerto Blest S.A., Catedral Turismo
Travesias fotográficas
7 Lagos car rental
Special thanks to Gustavo Galligaris, Sergio Juan Barbagallo, Mariana, Fernanda Arteche, Romina Cambiasso, señora Beatriz, Mariana Romero, Pedro, Daniel Esteban and Daniela Lacourt

Tijuca National Park:
IBAMA – Brazilian Institute for the Environment and Renewable Natural Resources
Rio de Janeiro City Government
Tijuca National Park Administration, Sônia L. Peixoto, Pedro da Cunha e Menezes, Flavio da Costa Balesdent and park rangers
Special thanks to Sônia L. Peixoto, Celso Junius, Bernardo Issa

La Amistad International Park (Panama):
ANAM – National Environmental Authority
La Amistad National Park Administration, Karin Rodrigues
Barú Volcano National Park Administration, Venero Marquinez
Hotel Los Quetzales, Carlos Alfaro
Special thanks to the park rangers of las nubens, Svein Vogt, Marta, Zulay and Bebel, Maria and Juan Fernandez

Galapagos National Park:
Galapagos National Park Service (SPNG) – Eliecer Cruz
Charles Darwin Foundation – Roslyn Cameron
Natura Foundation – Cecília Falconi and María Fernanda García
Lindblad Expeditions, Tom O'Brien
Ecoventura – Santiago Dunn
Special thanks to Tuy de Roy and Mason Fischer

Our supporters have played a fundamental role in this project, which involved a dream and faced many obstacles along the way. To them, our special thanks.

Suzano Papel e Celulose
Lan Chile
Advance Turismo
Nikon T. Tanaka
Ibama
Half Dome
Fujifilm do Brasil
Furtado & Schmidt

Turning an idea into a book entails considerable expense in human energy and also in natural resources such as paper, water, and ink, among others. There is an environmental cost involved in the process and for this reason we have taken some compensatory decisions in the realization of the project:

We have used Reciclato® paper produced by Suzano Papel e Celulose for the initial and final parts of the book.

Reciclato® is the first Brazilian offset paper to be produced on an industrial scale from 100% recycled material. Part of the income obtained from sales will be allocated to social and environmental projects operated by the Instituto Ecofuturo, a non-governmental organization set up by Suzano Papel e Celulose to promote education aimed at integrating men and nature. In addition to this, a part of the sales will be allocated to environmental projects.

This book has been printed using Reciclato 150 g/m² paper,
produced by Suzano Papel e Celulose,
and Magno Star and Magno Star Satin 170 g/m²,
produced by SAPPI, with a print run of 5,000 copies, by
Editora Gráficos Burti, in February 2005.